THE PSYCHE
TECHNICIAN'S
SPECTACLE

CASE STUDIES IN

SPACE TIME

Copyright © Dominic Haarhoff
1st edition, 2017

ISBN: 9780620748988

© Senoldo | Dreamstime.com
Cover: © http://i.imgur.com
Design: www.g3d.co.za
Set in Calibri pt 12

This book is not intended as a substitute for the medical advice of a psychiatrist. The reader should regularly consult a psychiatrist in matters relating to his/her health and particularly with respect to any symptoms that may require diagnosis or medical attention.

The information in this book is meant to supplement, not replace, proper psychological training. The authors and publisher advise readers to take full responsibility for their safety and know their limits.

Dedication

Sybil Barnett

(July 13, 2014)

Mary E. McBrine

(March 1, 2016)

Johannes Mati

(6 July, 2016)

From Alpha, The Earth and the waters beneath
through Homogeneity to Omega

Table of Contents

Illustration Index

Unless otherwise stated pictures are from Wikipedia and have a Creative Commons free arbitration license.

1

The Omniverse

The Omniverse concept has been notioned by many philosophical writers and fringe theorists alike, whom contend with string theory, as well as the esoteric science's of modern mystics from the last century. There are also some aspects mentioned throughout the various ancient Hindu scriptures. This synopsis will extrapolate upon the Wikipedia entry on the subject thereby setting the context for the following alternate interpretation.

The Omniverse is the cosmology of which centres upon the notion of an all-encompassing hyper-dimensional hub, by which all possible universes with all possible laws of physics occur. In the physical context, the limitation of the definition "universe" is that it only has one set of "physical laws and constants that govern it, whereas the Omniverse is expanded to include multiple sets of physical laws and constants, each expressed as a wholly or partially separate universe.

The later being universes which overlap into one another, like two TV channels being played simultaneously on the same TV, of which broadcast the exact same show, except each show has the potential to diverge into alternate possibilities of expression.

The Hierarchy within the Omniverse

- Universe: The inside description of a context that is relative in size/structure (attributes/modes) to the known universe that we inhabit. A Universe, also known as a

Cosmos, is a particular individual space-time organization with a specified number of dimensions of space and time and definite and specific laws of physics. Other Universes (other Cosmoses) may have different numbers of dimensions of space and time and different laws of physics than ones own Universe (Cosmos).

- Multiverse: The part of infinity that directly joins a given universe with all possible alternate configurations of that universe i.e. the field which spans a given cosm and all its subjective alternate cosms.

- Metaverse: In string theory, the part that is along with, after; over also denoting change in the multiverse that houses the branes or film that each universe is said to be attached to and hang like individual sheets in a hypermagnetic wave with rhythms of hypercosmic strings going up and down that has a third element causing up, down, backwards, forwards, motions inside the Xenoverse.

- Xenoverse: the unknown alien elements that are beyond and part of the metaverse and multiverse structure. Compared to a patchwork quilt hanging on a line to dry in space that is multivariate inside the Omniverse. While the Omniverse is said to be the outside ring beyond all that is known, the xenoverse on the other hand is the hypermacrocosm within the Omniverse that is unknown beyond the metaverse—the unknown sets of laws that govern how branes behave to create metaverses, the laws of which govern the creation of multiverses.

- Hyper verse: Multiple xenoverses, probably a quarter of a Omniverse which they are relative of [sort of like saying that the hyperverse is a harmonic membrane of xenoverses which contain the rest of the omniverse, so the other three-quarters exist within the hyperverse but are not really innately characterised by its laws directly, as it is more so akin to a field which holds the xenoverses together... The xenoverses then hold the metaverses together which in turn govern and arrange the multiverses various laws of physics (also known as codes of creation; infinity codes) Thus the omniverse holds the hyperverses within it of which span its whole volume but their laws only govern the arrangement of xenoverses, as such the laws that are attributed to the hyperverses are only a fraction of the total volume of that which they contain. So it is like levels within levels, each of which are governed by the level which proceeds them but in turn each level above has lesser relation t subsequent levels below the level that they govern directly

- Omniverse: All possible attributes and modes are in play, multiverses are categorized by the attributes/modes active in its child universes. Some or all possible modes of existence are actualized. If we take the point of origin as our being as a point in measurement, then we can generate the following hierarchy:

One can think of the omniverse as a tree structure: the omniverse is the trunk, the metaverse is the set of laws that govern the formation of branches, each multiverse is a branch, and each universe (cosmos) is a leaf. Alternatively, the omniverse can be illustrated as a forest in which a metaverse is the set of laws that govern the cosmic ecology that determines the distribution of trees in the forest, a multiverse is a tree in the forest, a universe as a branch on that tree, and all further branches and leaves are further subset horizons within that universe. [crosstalk]

Holistic Interpretation

The Omniverse is expanded into greater perspective by contending that space is fractal - holographic; holograms are pictures made of smaller replicas of themselves on a film strip, which are like a prism. When a laser is shone through the film strip the many small pictures all superimpose into the laser beam and are refocused as a single image upon a receptive surface.

To say the universe is holographic is to infer that cosmoses are made of subjective micro-cosmoses, like wise the micro-cosms are made of subsequent micro-cosms add infinitum i.e. the quantum foam, or rather macro-quantum foam (as above so below, as below so above). Each cosmos is defined as a static hypersphere whereby space implodes rather than explodes i.e. an Imploversial Scope.

Thus each level has just as much potential of complexity as the others, because space has infinite resolution, like pixels on a computer screen but an infinite number of sizes/levels of resolution all of which are accessibly reorganised into quantified cohesion at every possible level into infinity. So, essentially a universe/cosmos can sit in the palm of your hand and still be just as full of energy as your own objective universe.

Thus in contrast with the string theory approach of explaining the Omniverse, the holistic approach of Imploversial physics can yield a greater level of comprehension. This being because the quantum foam of any cosmos is made of subjective microcosms/quantum components, these components are repulsed by one another equally like a positively charged colloidal silver solution. Thus they are statically suspended by one another in space, but they each have a variate rate of depth charge (implosive expansion). This in turn creates contrast and clusters who's function is similar to a 3D grid of LED lights, whereby movement of particles as independent singular things is as illusory as LED lights consecutively switching on and off in progression producing the impression of a single light (or lights in unison) moving across the grid. This is called the scalar effect

Therefore from the macro-quantum perspective, microcosms arrange themselves into clusters – multiverses, which can be thought of as particles of matter in the quantum foam of a single cosmos within. Likewise a single cosmos can be thought of as containing the membranes of metaverses and xenoverses etc which are the various multidimensional fields that serve as nexus conjunctions for the transferral of holographic information i.e. scalarly morphing depth charge contrast between the quantum components (micro-multicosms) that comprise particle clusters in a single given cosmos. Thus a given cosmos also contains micro-versions of all the hyper-macrocosmic fields of the Omniverse, i.e. as above so below.

This means that all possible laws of physics can be exhibited within a given cosmos, it is only the accepted model of mainstream physics which defines the generally perceived limitations of an individuals everyday reality. Therefore the consensus reality is nothing more than a widely held belief which filters out the awareness of other laws of physics at play.

The Quantum Measurement Problem

In quantum physics, all measurements require a conscious observer. Yet to explain consciousness, science theorises that it involves the interplay of smaller and smaller particles (neurons, molecules, atoms, quarks, photons, electrons, etc.) The existence and state of these discrete elements is dependent on quantum theory which requires a conscious observer to collapse the equation. All the whilst the actual essence of the conscious observer or "consciousness" its origin of eminence, remains

11

unexplained (i.e. "the black box ".)

Since consciousness is unexplained by physical theory but remains present in our understanding of physical theory, it is the mother of all things (even the unconscious since it cannot generate the "conscious") The measurement problem is resolved by unifying the observer and observed into a single system that is not compatible with traditional reductionist science - breaking things down to understand them)

The unification is achieved by accepting that there is only a single universal consciousness that all seemingly separate individuations are a part of. ("God" or whatever one so wishes to call it).

Quantum theory cannot resolve the measurement problem because it is a metaphysical issue. However what it does reveal, is that science is still unable to falsify the perspective of mystics regarding their take on the deep mystery surrounding existence and the purpose of individuated experience.

1. the primary universe (in Flux)
2. the multiverse (shell 3) Celtic
3. the metaverse (shell 4) Marvel
4. our location in space-time,
5. our location our universe (shell 5) Islam
6. the xenoverse, (shell 6) Marvel
7. the hyperverse, (shell 7) Business
8. the omniverse. (shell 8) Science

[crosstalk]

2

Universal Scale down to DNA

In early 2011 I set myself up in a private practise as a Lay counsellor with just myself. Later I would meet others one brave Irish plastic surgeon and minister, Mary an educator, night nurse and medical supervisor Sybil and one worn torn previously lame physiotherapist. Over the past three years I thing we rotated through 36 physiotherapists. Together we would brave 13 income areas around the world and know much acclaim and success. We would also face great hardships.

I started out knowing very little about the discipline of Psychology to mastering its concept when it came to clearing myself and living conditions and malaise of mental health conditions. In the beginning the most we could handle was 108 for each client / year.

Eventually we invented a system where we each could facilitate healing very quickly from the Femto (Universal) scale down to Lambda DNA by combining teachings and developing mechanisms for healing the psyche of the client. We could handle up to 2048 relinquished relieved per session.

My intention to write this book came in 2015-6 after grieving and mourning the passing of more that 3 mentors. Beloved in the fields that we shared using the same techniques we had developed. I was left with a pack of cards of sacred geometry and a partially worked manuscript for the book you are reading now and Sacred Shapes for Children. And here I am now in 2017 where I am, waiting, alone once again.

I have many values that I inherit from my grand parents and their four families that enable me to ground and make practical the morals (like the 10 commandments) and when I do them well, yes I develop or cultivate more virtues (like the 8 noble truths of Buddhism). My essential values are just for today not to anger, not to worry, to give thanks to every living thing and to honour my parents and elders (tradition) I also have the values of cultivating intellect, integrity, sanity and faith in my family as well. We have the values of musicality, justice and work ethic. When I did genealogical research into my family and from spiritual sources when there was little evidence I found the family to often be the role of priests, carpenters, merchants, judges, wardens and doctors. I do feel sometimes like I'm standing on the shoulders of giants in this way when I feel their support in my work.

One of three different ways or set of beliefs that that I use everyday in my counselling practice the biological model. In my lay counselling / life coaching practice I assist others to overcome their emotional and mental inheritance and other facets of tans-generational transference or otherwise deal with the legacy a person's parents has left them. According to this belief we assume that suffering has occurred and led to an impact on the DNA or stress or trauma on any of the levels science has shown to exist. There is much research to support that parenting styles, heavy metals and diet have an impact on the DNA. When I assist someone to become self actualized from I biological perspective I see that they get better at reaching a higher brainwave state over time. And their focus moves up Maslow's pyramid from physical to social networking to metaphysical needs.

As the therapy process continues people often after having become self actualized and go through changes in their long term beliefs. Once they manage to see their dreams during the day under due to therapy having lifted them from their mental and emotional worries. they wonder about how they fit into the collective unconscious which occur when one attains a gold color in their minds eye. They are no longer just an individual making a living they are now drifting from white in or the state of delta in their minds eye during the day to bringing down something from their great unconscious. There a re lovely models to explain this in African traditions like the reality of Ubuntu and of course Jungian psychoanalysis and certainly from a certain angle science proves some of this too when it says there are 12 dimensions folded into space- time. During my therapy process sometimes I like to imagine that different factors of our energetic anatomy live of these levels. I have developed over time ways of assisting people to clear their id, super ego, conscience and other collectors of energetic debt that affect their relationship with the great elsewhere the grand sentience behind everything in existence. So from a certain angle I like to see tat we have multidimensional parts to our psychic anatomy.

So aside from the hierarchical (angelic) model found in Christianity I sometimes like looking at things from a Oneness perspective. Thus the client could see that their archangel is helping them through a tough time and later on can see it from a oneness perspective where a more functioning part of their ought to be self was assisting them. I often wonder where the boundary between us and another person should be drawn. From one perspective we are connected on the subatomic level and the other we are separate or seem to be individuals. There is also much debate about what happens to our sentience after we pass on. I prefer in therapy to give clients facts and experiences and not tell them what to think. I like assisting people to form their own opinions about matters of faith. As you see here I have at least three systems of meaning I call on. Not one of them can deal with 100% of the evidence I have collected over the years as they are based on different premises.

One modality I use often in my therapeutic work is past life regression where a client might request it or go back to a moment in time that occurred before their present life. This often happens to people who I have done work with to counsel them through their present life. Some like to look at things depending on their beliefs that their soul has been through many lives. One can also see these encounters metaphorically and another way is to see this as being sins of the father that become sins of the son. For me in my time I've seen enough evidence for myself to understand that out consciousness does survive before and persists after death.

E.g. I once has a client who could remember 7 years of his life with his mother before he was born and often assist in lay counselling people to deal with the prospect of surviving death. In this belief system we prepare over many lives to become enlightened and achieve mukti or cessation of rebirth. My understanding is that we join up with the cosmic background radiation of or music of existence.

The Reiki universal light beams reflect through the prism of the psyche like a rainbow. For the Thanatos the purple ray. Super ego the Silver ray. Nadis are released via the gold rey. The mood level is relinquished relieved through the White ray. The Mental level via the Green rey. The chemical level through the Red rey and the DNA level through the Yellow ray. The lepton level through the White-Blue rey. By mixing the above colours one can treat a wide spectrum of energetic dis-ease. Standard Psychiatry models use pills to treat 12 levels from DNA up to the Thanatos. When I work on a client I clear 21 levels to allow them to become Super self Actualised.

My Linage is as folllows: c/o Dr Usui > Dr Hayashi > Mrs Takata > Beth Gray > Vicci Grant >> Shaman Elder Maggie Wahls – Life Healing Community Winona, Mo USA

Bosonic String Theory

In 1974, Claude Lovelace discovered that bosonic string theory could only be physically consistent if it were formulated in 25 spatial dimensions, but so far as anyone knows, we only have three spatial dimensions!

Dimensions are the pieces of information needed to determine a precise point in space. (Dimensions are generally thought of in terms of up/down, left/right, forward/backward.)

Relativity treats space and time as a continuum of coordinates, so this means that the universe has a total of 26 dimensions in string theory, as opposed to the four dimensions it possesses under Einstein's special and general relativity theories.

Einstein's relativity has three spatial dimensions and one time dimension because those are the conditions used to create the theory. He didn't begin working on relativity and just happen to stumble upon three spatial dimensions, but rather intentionally built it into the theory from the beginning. If he'd wanted a 2-dimensional or 5-dimensional relativity, he could have built the theory to work in those dimensions.

With bosonic string theory, the equations actually demanded a certain number of dimensions to be mathematically consistent. The theory falls apart in any other number of dimensions!

The reason for extra dimensions

The reason for these extra dimensions can be seen by analogy. Consider a long, loose spring (like a Slinky), which is flexible and elastic, similar to the strings of string theory. If you lay the spring in a straight line flat on the floor and pull it outward, waves move along the length of the spring. These are called longitudinal waves and are similar to the way sound waves move through the air. The key thing is that these waves, or vibrations, move only back and forth along the length of the spring. In other words, they're 1-dimensional waves.

Now imagine that the spring stays on the floor, but someone holds each end. Each person can move the ends of the spring anywhere they want, so long as it stays on the floor. They can move it left and right, or back and forth, or some combination of the two. As the ends of the spring move in this way, the waves that are generated require two dimensions to describe the motion.

Finally, imagine that each person has an end of the spring but can move it anywhere — left or right, back or forth, and up or down. The waves generated by the spring require three dimensions to explain the motion. Trying to use 2-dimensional or 1-dimensional equations to explain the motion wouldn't make sense.

In an analogous way, bosonic string theory required 25 spatial dimensions so the symmetries of the strings could be fully consistent. (Conformal symmetry is the exact name of the type of symmetry in string theory that requires this number of dimensions.)

If the physicists left out any of those dimensions, it made about as much sense as trying to analyze the 3-dimensional spring in only one dimension . . . which is to say, none at all.

Dealing with the Extra Dimensions

The physical conception of these extra dimensions was (and still is) the hardest part of the theory to comprehend. Everyone can understand three spatial dimensions and a time dimension. Given a latitude, longitude, altitude, and time, two people can meet anywhere on the planet. You can measure height, width, and length, and you experience the passage of time, so you have a regular familiarity with what those dimensions represent.

What about the other 22 spatial dimensions? It was clear that these dimensions had to be hidden somehow. The Kaluza-Klein theory predicted that extra dimensions were rolled up, but rolling them up in precisely the right way to achieve results that made sense was difficult. This was achieved for string theory in the mid-1980s through the use of Calabi-Yau manifolds.

No one has any direct experience with these strange other dimensions. For the idea to come out of the symmetry relationships associated with a relatively obscure new theoretical physics conjecture certainly didn't offer much motivation for physicists to accept it. And for more than a decade, most physicists didn't.

Mirror symmetry

The Clebsch cubic is an example of a kind of geometric object called an algebraic variety. A classical result of enumerative geometry states that there are exactly 27 straight lines that lie entirely on this surface. The Clebsch cube can condense and compact a reduced a clients share of the universe we live in. [crosstalk]

After Calabi–Yau manifolds had entered physics as a way to compactify extra dimensions in string theory, many physicists began studying these manifolds. In the late 1980s, several physicists noticed that given such a compactification of string theory, it is not possible to reconstruct uniquely a corresponding Calabi–Yau manifold.[96] Instead, two different versions of string theory, type IIA and type IIB, can be compactified on completely different Calabi–Yau manifolds giving rise to the same physics. In this situation, the manifolds are called mirror manifolds, and the relationship between the two physical theories is called mirror symmetry.

Regardless of whether [Calabi–Yau conscience compactifications] of string theory provide a correct description of nature, the existence of the mirror duality between different string theories has significant mathematical consequences. The Calabi–Yau manifolds used in string theory are of interest in pure mathematics, and mirror symmetry allows mathematicians to solve problems in enumerative geometry, a branch of mathematics concerned with counting the numbers of solutions to geometric questions. Once riding in a Calabi–Yau manifold one can go on an adventure in the eternity outside the omni-verse. Since sentience can be folded safely inside each nested design.

On ones way back whether you travelled in your soul or self or spirit body or folded sentience you punch through the punch card system and land back with a matching pattern to what ails your psyche. This technique is used especially in clearing hard dot reach expenses inside the omniverse.

Generalizing this problem, one can ask how many lines can be drawn on a quintic Calabi–Yau manifold, such as the one illustrated above, which is defined by a polynomial of degree five. This problem was solved by the nineteenth-century German mathematician Hermann Schubert, who found that there are exactly 2,875 such lines. In 1986, geometer Sheldon Katz proved that the number of curves, such as circles, that are defined by polynomials of degree two and lie entirely in the quintic is 609,250.

By the year 1991, most of the classical problems of enumerative geometry had been solved and interest in enumerative geometry had begun to diminish.[101] The field was reinvigorated in May

1991 when physicists Philip Candelas, Xenia de la Ossa, Paul Green, and Linda Parks showed that mirror symmetry could be used to translate difficult mathematical questions about one Calabi–Yau manifold into easier questions about its mirror.

In particular, they used mirror symmetry to show that a six-dimensional Calabi–Yau manifold can contain exactly 317,206,375 curves of degree three. In addition to counting degree-three curves, Candelas and his collaborators obtained a number of more general results for counting rational curves which went far beyond the results obtained by mathematicians.

Examples of wrapping you consciousness and travelling in Calabi Yua manifolds. 1) Zen Journey, 2) Islamic cleansing, 3) Scientific catharsis clean-up, down, left and right.

1.   2.

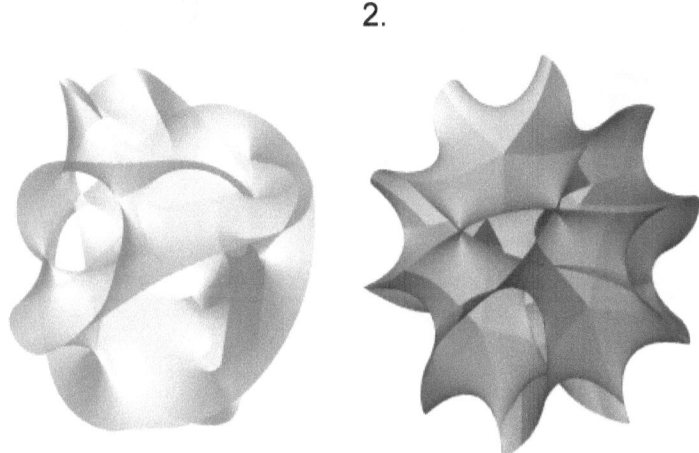

Representations of 2d-slices of Calabi-Yau quintic manifolds

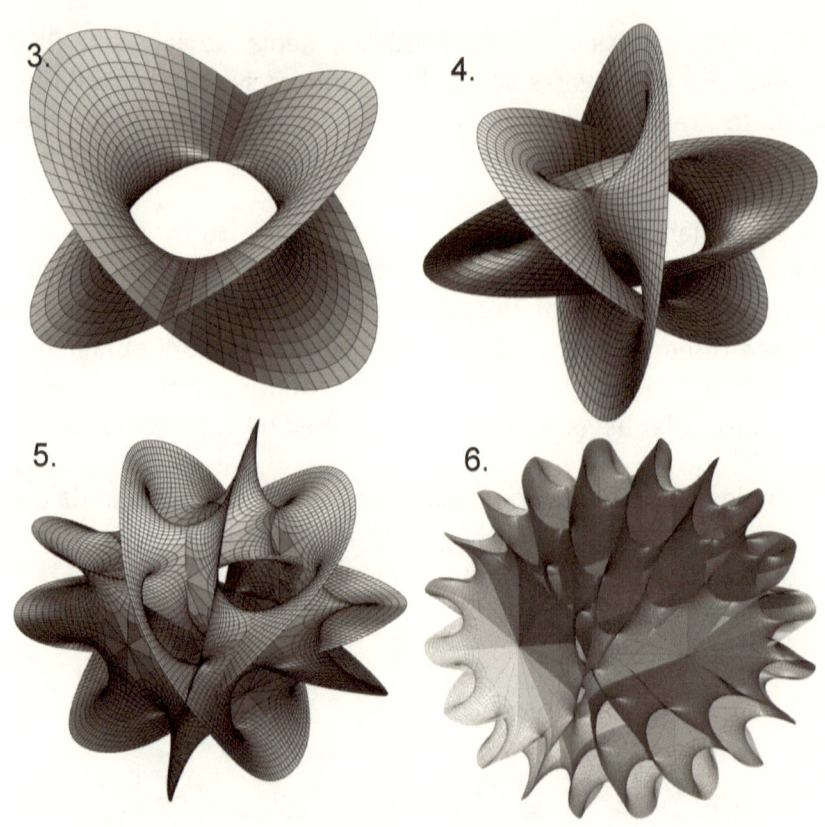

3

Geometric forms of the Psyche

There are fine variety of geometries used to clear a built up psyche object that is active In real space. Designed to relinquish relieve a person after noting the nature of the Dukkha (Pāli; Sanskrit: [crosstalk]) an important Buddhist concept, commonly translated as "suffering", "pain". The psyche can hold around 10 psyche objects at one time.

The facilitator needs to ask probing questions and write-up the qualities encountered during free association and administering the talking cure. They may need to ask the client is they will give committal or permission to release their sadness and in the beginning of therapy.

You may need to ask one by one and only cleanse what they allow as each person has a limit of how much jhanna, karma, sin or debt they are able to process over a week. E.g. to clear a 4-cube picture a blank one in your mind that is fixed and simply write the ½ opposite of a clients affect on each side and inside and pass it on to bring it to deliverance.

The method is to in the beginning of you application of this technique. Imagine or visualise the exact hyper cube (for 4th dimensional cleansing) and fill in on paper what you processed that week. 1 opposite – ego, ½ opposite 4D, 5D, ¼ opposite 6D, 7D. 1/8 opposite 8D, 9D, 1/16 opposite 10D, 11, 1/32 opposite 12D.

Psyche technician tool kit:

- Crystalline mirror (Scrying and vision)
- Sledge hammer (travel up the spire)
- trebuchet (long distance dream sending)
- tesseract (pyramid healing – cf a surgeons scalpel)

2048 Dimension - Chrisencube

Christianity has many symbolic connections to the [Flower of Life]. Most notably, the Seed of Life and components within the Seed of Life have strong Christian meaning to them. Such components are the Spherical Octahedron, Vesica Piscis, Tripod of Life, and Tree of Life (Kabbalah). Also the symbol of Metatron's Cube is delineated by a component of the Flower of Life and has appeared in Christian art.

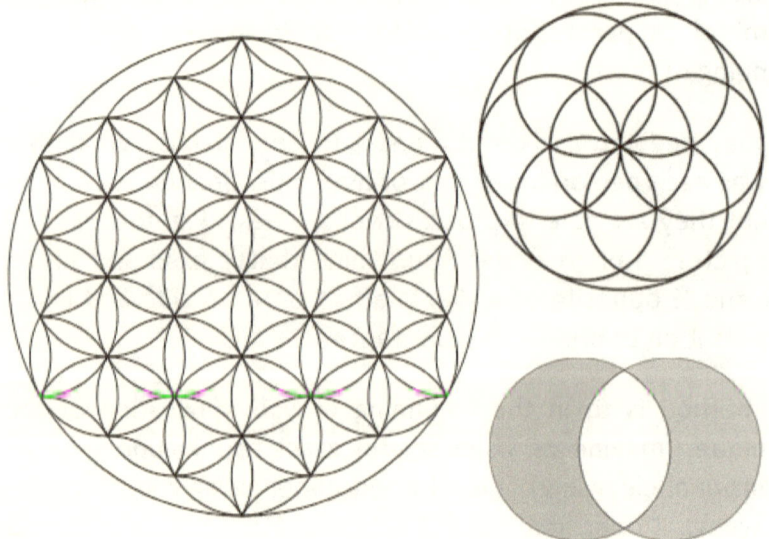

Illustration 1: Dimiletetracontaoctagon = 2048-gon

68 flowers of life is enough to reduce a packed chirsen-cube

999th Dimension - Double torus

The 999-gon Used to access the needs of ones children:

% Clearance depends on number of children. So with 2 children each has 50 % of the vote – 4 children each a 25% of the vote and so on.

Illustration 2: 999gon

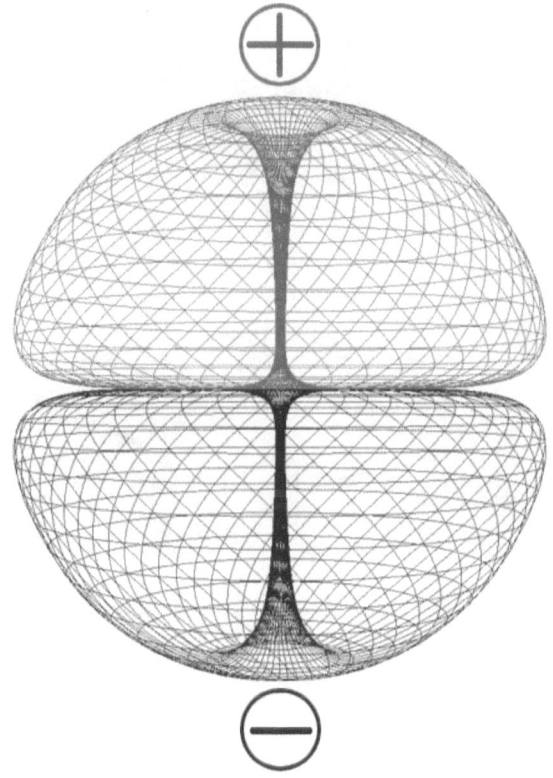

12th dimension – Thanatos

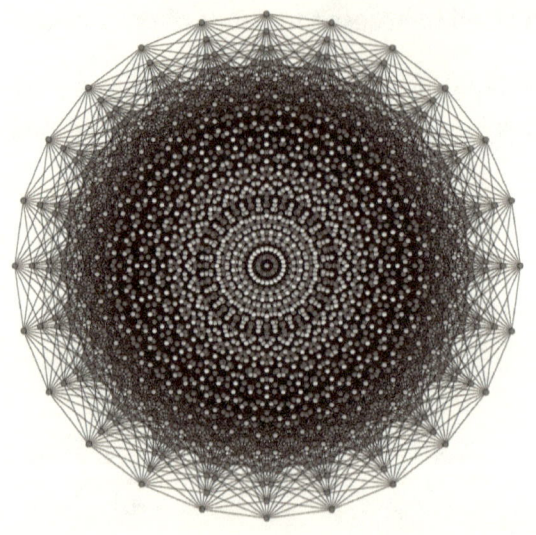

12-D cube

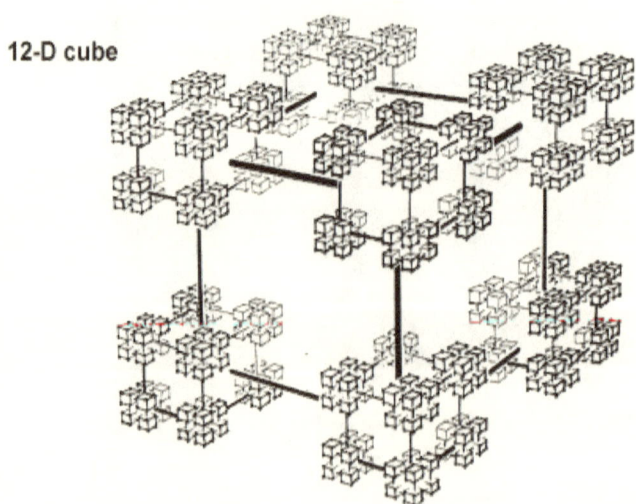

Illustration 3: 12D hypercube

Dancing with death

Before I had all the fancy methods and modalities I now enjoy today all I could do in the beginning is emote with the client, stay angry when they are angry and hope that the talking cure would be enough. I met Joel, a Pranic healer, at a workshop I was offering on Shakti and Shiva healing in my home town.

Joel was complaining about life times and lifetimes of abuse committed by him and his farther. And incessant reincarnation the same family. When I began to treat him with body work in my first contact session he immediately relaxed and after 30 minuets of scenes from the past and looked re soundly better.

During the talking part we spoke about his dreams of Ancient Egypt and work at a funeral parlour. Future treatments would eventually allow Joel to become super actualised. During the next few sessions. Joel is carried through whiplash and being wasted through the Thanatos in his dreams and psyche 33 times and then something unexpected happened. A bright silver light flashed past.

His thanatos was dusted and relinquished relieved. The Id followed. It was as if someone from above had just teaked something of brilliance inside. I would later learn after more studies of sacred geometry that merely watching the psyche object from all angles would be enough to make it go to dust and after this one would not see this for good.

However despite making it through what must be heaps or onmiverses of debt. We were still along way, we would go through the entire solar system before finding the forgiveness in Joel's male line and welcoming in the women.

Dominic: Good morning to you.

Joel: How has your week gone this week. Good. Mine too.

  Dominic OK good song. Last time. We went through an epic clearing that I think is the last. You're going to need. And we had we went to your transference you saw ancestry you had the lineage scrum and you saw everybody at the end because your grandfather. And your father. And then nice people popped up. Were willing to support you. You got a lot of support out of that session which was which I think carried you over the limit that you needed. To release all sadness.

Joel: Yeah. Yah.

Dominic: So we were a big journey with a bit ahead of everything big time.

Dominic: Okay so tell me about your week.

Joel: Um. Sure lets see. Energy has been good. I had a few clients this week. Not too many. We are re-staining the outside of the House. So we're house. With brushes. We will do the same. Sleeping good. At some point we felt into an energy in Egypt. It will save you some time over the weekend. I did really fill the place. In little of day. I finally tuned into it. That was on Monday and he shook his head. If we go those donated to people that they do a group look carefully chipped. She will still be here just kind of shining where the grid. Wants that energy needs to go. Go. To the head of one of funeral dream that I remember. Here. I think it really should be about being.

Joel: Some place to access tying or at this time a tying of the dream.

Joel: In a nutshell. OK good.

Dominic: OK good. Song. We're going to do a puzzle.

Dominic: OK. So today we're going to do with shadows. And we're dealing with. Your grandmother your dad.

Dominic: And I'm wondering if you feel. If you if you feel up to doing a scientific journey before that or if you want to do anything different. To prepare you for that. You've had a whole week to think about everything and I've given you many things to think about and to watch and many dreams. When you can. There's this tremendous amount of love for you in the family. You founded and you are generated from. The final release for your journey lies with your dad and that means that you need to get up the courage to. Meet him in one of the tunnels where he is and I will get you there as a Counsellor on a visa.

Dominic: If you want to see him. And be released from him for good. On one side positive that happens to him goes well. On a downside it might reawaken many small services for you but you seem to be over it for long enough to be able to withstand seeing him maybe once this is the same as meeting him in his worst condition. And you must have some ethical challenges to work out and you will have to say whatever you feel it feels right to you if you want to. Scold him if he won't accept him marginally big time. If you want to say they can give him more food where he is if they can give him a nice day or nice routine to follow. He will soon. You will see only see where he's living.

Dominic: And it's not the nicest conditions of all that's the reason being that's no condition at all. So you will see when I get there and when preening for a week to get to this place.

Dominic: You just go down the tunnel. The benefit is at least that he's in a respectable condition. Everyone around him is golden. He doesn't feel like running in his duties because he still feel sad because he's falling to you and to your wellness.

Dominic: So your dreams have changed for the last three years. I've known you. And you're not getting the funeral dreams occasionally and you will get something more joyful and joyous. I think from this activity alone you will be some there will be some boost of a Sunday as of Friday. And I think it's Monday the scientists show the most or not Wednesday recompenses traditions do you honour that you have different genres and different traditions that you're on it would be and which day would be reserved like that but one of the days must be because of his we not two of those then it was a nice day.

Dominic: That could work out quite well. So there are many possibilities of what could happen out of this process but I will take you down there as long as long as you need to feel strong if you feel that you can get you need to get out. I will escort you away.

Joel: Yep I'm good and there's nothing he can do to me there. Thank you.

Dominic: OK. Good. So if you would like to I would like you to close your eyes. We'd be happy to have you listening on with the video you vision and then this thing. Or would you want to just deal with listening fool for the first. Part of the hour. OK. We're going to meet him at your best to go.

OK. When you open your eyes inside can you tell me what you see and what you hear.

Joel: We're very. Not hearing anything yet.

Joel: We're seeing some still adjusting to the place I feel some heaviness in my heart. This place is.

Dominic: You feel any sensations inside. By sending zooming in plunging to. Whatever you might be doing then. You are going gently downhill does you feel like it there's a sharp drop where are living in and it's almost like someone is trying to pull you.

Joel: Receive just as it would rather be close. To energies This is either that or you are most likely to say OK good meditate some of this would be somewhat easy and you won't notice it too easily.

Joel: It's. Pretty interesting. You know much later here.

Joel: I'm sensing two rocking chairs. He's the one in the other.

Joel: He was. A big rocker.

Dominic:Joel What would you like to share with me about what you are experiencing.

Joel: He is apologizing to me you're not really. Hearing. Words clearly feels like. He's trying to apologize for his actions. He's talking.

Dominic: So today you have proven your strength. You are proving your mastery. Of your own fear and of your own. Culpability culpability is a very interesting one this means that. You have to agree that he's put you through many different situations in psychological terror.

Dominic: You had to do work with his outs for his house and it was not easy. The maximum. Killing to belong and lying to be along and even lying to him about how much time elapsed during the. Encounters. So it's not easy. You have to admit that you've actually done many things. To try to make the pain go away and when you when you succeeded you were both noble and true.

Dominic:] It was. How I know you.

Dominic: Shame and guilt. to go is not an easy thing. It really isn't.

Dominic: Shame and guilt is the Opposite of joy. So if you find yourself in a Joy then I'm going to send you these yellow lights to help you see. It's very daunting. He's chosen to be in a cave.

Dominic: Had a preview coming into this. Wonder how beautiful the tunnel is the golden tunnel. It is nice that people agree that they give permission for them to be treated. It's not easy otherwise. But he's dead on a long time so it's possible he's running out of time has passed. For him it's been that he says that he minutes but it's actually been a few you know been a few hundred years.

And what's going on live in is the sunset it's been 2050 years is not easy. You just have to live in a cave and don't want to give him on insurance. So there's a lot that you feel that you need to please just hop on top of the glow like a glove you know numbers and then you will feel joyful which is the opposite of shame and guilt.

Dominic: He says please offered him to in a way that he will understand that someone has come to him today. I don't know anything that they feel that their life is going on and he will know that your life is going on. Since he died. So it's not it's not easy to tell. He will not know that in this form he will not know how long it's been for you and in the darkness. No one will tell. But at some point you could say it's been a long time.

Dominic: Make what ever statement you want to make to him in private. And then let me know when you are ready to continue. With switching more senses on and I think it's easier just to talk in the darkness it's easier to use this voice in the dark you get used to the joy lights and then the joy- contentment nimbus.

Joel: SO I'm forgiving him. Indenting to release him from all space and time. them to release him for us. Oh space and time so energy no longer has any effect on mine.

Dominic: At this point. We were wondering about. If you want to put. A light on. If you want to see. His face. What are you going to do if you really wanted to go on. You seem to be in. A small little compartment like a track or a bird or a. Small little vehicle. And he's stuck in the in something like a time rift somewhere. And he's been there a long time. He doesn't have consciousness of time.

Dominic: The law of. When you get something you are bound to get something in return. But that's the law of common forces. It means that to every one it falls onto like one with an arrow that someone in judgement you think is behind back and you've had this before on the playground you've got to one and maybe the physics room.

Dominic: One of the poses was the reaction. Was negative and some positive. There's nothing like good karma but at least you have something like lot of positive causes which is shows that. At least if you give to me you will get in the one that you seek. Have you done everything that you want to do today. That's what you want to do and I will bring you up and I'll hit you up and I'll change aspect again.

Dominic: And I think it would help his development would help police 438 people 432 whatever that number was run the 400 is right there that he has harmed. Would help to the management to sustain. Put you against your will. And it seems that you are ready which is why I spoke again at that point.

Joel: I'd be happy to do what ever I can to let them get unstuck.

Dominic: Your heart is ready for a big change your conscious mind is a bit. Pessimistic as to how it's going to be. And your self is kind of neutral at this point they're not showing an opinion either way. So you've got your soul in a toy truck and you are on voyager mile seven

Joel: And I just indented to turn the light on. And you then if you want to see the launch is it to reach for the light and then put on.

Dominic: You turn away good.

Joel: So I intend for his soul to go on. For him to do work he's been putting off. She's to. Receive whatever assistance he can receive. To facilitate his growth.

Dominic: OK everything around do is set up that he if he chooses to accept help. Should. And if this is 50 percent complete he would he would be able to. See the light at a later stage. By 50 percent or more of his of his victims had to vote to vote for him that he gets wants knowledge and that he has a change of heart about accepting help from anybody else.

Dominic: Can you put the light on.

Joel: I guess I already intended to.

Dominic: Radius is 30 meters that should give you some vision. OK what do you see out there.

Joel: I guess I can describe it as him out there I space in a spider web.

Dominic: Tell. Me while we are busy and have not started and ask you a lot of questions about your family and your brothers and sisters and family and close friends when you're young but you've had many people who work with you over many years it's not easy. Is there someone who is who is who would be better to ask them to offer him something warm some clothes or a robe maybe something to do.

Joel: One of his dear friends passed recently . That's yours. It is. OK.

Dominic:At this point I feel that he's not he's getting he's 50 percent. He's permissive but he's not really getting encouraged. And the freewill choice is has to stick around where he is or to grow and develop and be happy and whole.

Dominic: OK his friend Mr. Wheat is in a vastly different place. Is it's possible for him to be here. Maybe next time we meet. [crosstalk]

Joel: OK.

Dominic: To have his assistance with the matter. What he was willing to do for the man. He doesn't have the same training but he's quite fine. This school is being run by a nice eternity afterlife warden who lives at this place who is able to handle requests in this quarter. It's quite an honour. It's stuck in a small uneventful place in the universe. One of the universes is the right next to ours and it is where he is and life is continuing on very happily.

And then as it explores visit occasionally and they found this place a while ago. But nobody was really willing to leave. And if you just stayed stuck there for a long time. And I was just inches from the tenant during the Super 8 which is where we could go to as a journey to find truth to find your truth to find your Grail to find your eternity clothes then you can be on a more even footing with your wife. She has 25 eternity robes. You have one with two black and red.

Dominic: An you are very much a middle world man you see that you see that you see the silver lining on the people. As you know your wife is somatic paraplegic so she doesn't feel the body too easily and she doesn't look like her. She doesn't use her hands that she eats and many many things that she doesn't have physical bodily sensation that much.

Dominic: And instead of that she has a green heart that you can see the world map and you can move on the map and move around and do environmental missions and what the consulars rather than this complete physical people too much. It's been a whole life of trying to figure out how to see the physical world and physics behind that and it was actually the opposite basically to be that. Another person you know had this for many many years and you know I've had a supportive role in making sure that you could give her a talk which needs to be in the hands of mealtimes and the many small things that by dexterity that you haven't grasped.

Dominic: But. I want to take you, Joel, to eternity and I wonder I think you've done enough for me to do this to you. I mean leave the current scene where he is but he's not going to budge more than you've given to me a lot of warmth and you going to the many joy packets of joy and positive parts of your nimbus and something to be happy about and he looks a bit better.

Joel: My brother, Craig, has also passed over. So I ask him to be there for our father.

Dominic: OK. So we're going to go on a journey for the next 45 minutes. I'm going to spend the extra 15 minutes with you to check you into eternity have some better activities then come out. It happens to be exam time season and you've prepared quite well and I'll give you many modules and books over the past few years. And you received them and they helped you with your life but now it's intended for eternity. Was that was kind of like. I want to share it with you. Not like this you know not leave the space in a 10 back to where we are to turn them all. In the town watch you know back in your body.

You could have many choices but there is war and training up for grabs. Wardens are people who deal with people like potentiality people like they do on Earth wardens and intended to deal with dreaming reprogramming of. Plastic surgery and entities and beings. We've seen a few beings on the roof. And the cost has come in and they top the scent and they don't seem to get more physical than that. But they enter reality into any sort of weekly and then about two three to seven point you'll be able to give. Fairly nice work on rounds and faces which means that you can do. The right distance from nurses from people who like myself. I'm quite I'm sorry. I'm very good with. Hitting people. I'm not so good at healing.

Dominic: Eternity people are not so good at that but it's in shape shifting and you might enjoy both a bit of transformation means. So sorting it means. Something different to do that. And it's the exams are today and I think this would help you a lot to understand how you're going to get clearance of. You first got to get your stuff and then you wanted to meet your dad and then you got to a point where. You might feel that in a third world you can meet with many other of your friends and not that you care enough you can go to the afterlife and you can finally get switched on for that with me. And I don't have the time to spend with you know only have. 15 minutes that I can climb with you to.

Dominic: Have you get you don't get that done it and by next week you will have the you do the exam and then after that I can show you two. You have a purpose to be there in the afterlife so that you can cross over and do something that and then you can do you can meet. Gather your brother Craig and everybody else that you need in a party and plan a plan a plan something for your dad to find to give him the choice and opportunity.

Dominic: To find solace. OK. So we're going to take you I'm going to take you briefly to. Take you to the tunnel and I'm going to see if you want to go right through and then you go to him from a Counsellor and you will get a package from them. And then I will leave you for a week and we'll come back in a week's time to plan your dad's. Intervention.

And you see outer space. Can you see that. Can you feel like that tunnel is pulling you backwards. Which is your time that you used to expedite life that monitors your progress and success attributes skills.

Dominic: And good luck. That's good. And then you've done an entirety of your. Lifetime's time getting along. Does it feel to you. Does it feel different to you and your body.

Joel: Yes.

Dominic:] The Reunion is going to be a big one next week and it's going to you're going to see many people in the act live up to that with the maximum 133 per person. That I. Think. Nobody needs more than addition and that and they get everything they need to continue with their life. They'll either hit on the Counsellors and possibly can put orders on the earth and sing with little many jobs to do that outside of eternity inside the omniverse inside the universe to deal with. Where we live. You can do this to explore a connection to something bigger than yourself - which is something that you've missed for a long time.

The gods you to many places and you go to many events but you need something and you can do yourself to assure you that you have the right orders for you to get. Great.

Joel: That. Is true.

Dominic: Spoken the truth doesn't always sound like what it is but it can be translated in many languages in many ways. You are going back a tunnel and you meet somebody you meet your Counselor. At the end of the tunnel that we never do. So we have about 10 minutes left. To do this and then next week we'll stop the Trinity mission to explore the like for you to see what's who's on there.

Dominic: To arrange an intervention for your dad. That will be a wonderful thing and then you leave it at that and then we could continue positively for quite some time if you wanted to. Then once a month would it be once every two months. As long as needs support for me to understand the aftermath and to see these things you need me and then you got a wonderful wife who shows people who love every single day and then not always do this and this is look at the universe and feel happy to see such a joyful thing.

Dominic: In the cosmic objects many things but you have a very lovely life in the whole world with. Wings and trimmings and many things so you have belief and you know that it's something that you just need a constant standing with if you will need something and I'm not sure what you need to cancel your concert I and many other people in the afterlife will inform me of many things and you will see that you have.

Dominic: Something to do with your plans. Wardens are very in tune with people's psyches you can eventually put your hands to decide their head roll under the under their skin. And then you get to touch their psyche and touch me just fine shooting it in.

You can't touch any of their parts and interact with them. More than psychology's more than counsellors more than teachers do with creating schemas. Can you see anybody at the top. How can you hear them.

Joel: sensing a being there Being there and possibly turn around. I'm not conscious that is how I would describe it.

Dominic: You feeling this present would you like to listen.

Joel: Sure.

Dominic: Do you have any instructions for Gregory because that's his name. Did you even that tell him that not to talk to you at some point when you were younger.

Joel: I have no idea. Right.

 Dominic: You can talk to me a little bit and put the phone on the phone on something listen to him and. Say if you wanted to be private you want discussion with him like a confessional like better come that is this and this and yes my child and does do that. I don't listen to anyone or was you were with him for you. Can you see him he's flashing pictures of the tunnel can you see that.

Joel: Um ha

Dominic: What is that picture like if you want to use it. It's in some language in my language what do you want to read it. What would you do with it. Not to share that with me would you. You don't need to. You don't want to just say that you should. That's helping you to understand where you are.

Joel: This is a kind of like a sunset. Overlooking the water of the ocean and mountain in the background. Is always encouraged. I guess I realize that he helped me in all ways that he could.

[I use the purple ray of Reiki to treat Joels Thanatos.]

Dominic: OK. You have a ticket to eternity for next time. Time 9:00 a.m. on Friday and next week.

Joel: It's been a pleasure.

Dominic: Thank you. Thank you.

[I give back what is theirs

I take only what is mine

I am love.]

11th dimension – Id

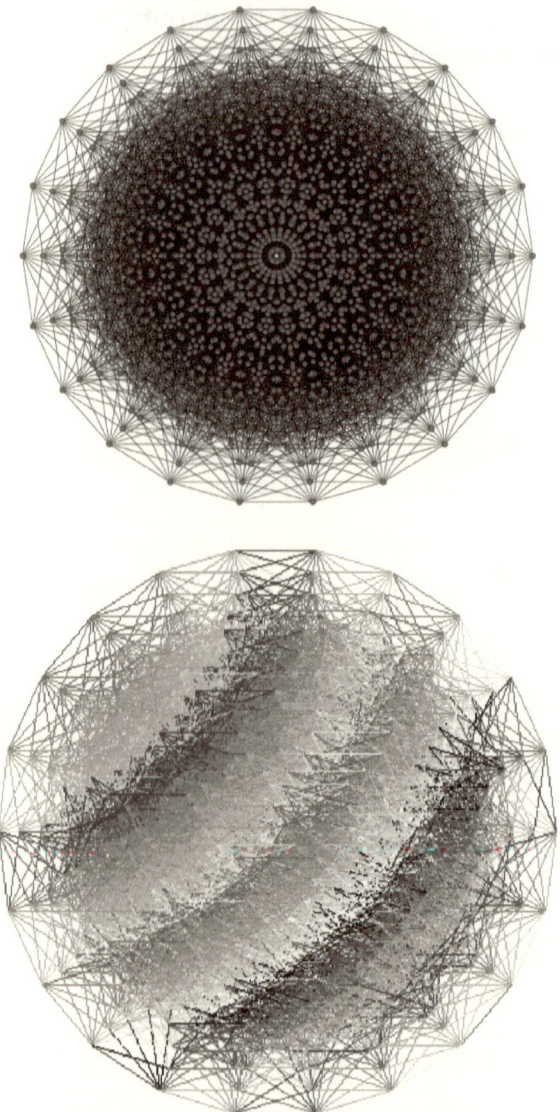

Illustration 4: 11D hypercube (11-gon)

Id cleansing

I first met with Frederica when she wanted some shamanistic work from me to open her senses. A number of months later I was contacted by her again after she had received corrective foot surgery. And we spent time relinquish relieving dusting and flattening her super ego – judgement process of the psyche. Now we wept forth the id.

[While I talk to her I treat her with the Blue Red and Yellow Reiki rays that mix together to form a dark Brown colour]

Frederica: Good morning. Yeah. There we go.

Dominic: Good to see you looking bright bright as new.

Frederica: Yes. Well as you noted.

Dominic: So for someone who's just recently had surgery on two feet you know moving around and doing all these things and you know every time seems to be much faster than it. Could have taken.

Frederica: What was crying me was the nature of it. And the beauty of nature of us. And me by the ocean. And my heart hit on me and I'm just not being grounded here.

Frederica: Yes you were right. I've done my journey. Man. Why I'm turning over. So. That you're so good in a nearly.

Dominic: When I last talked to you, you were sitting. You couldn't. Walk as much and you couldn't do as much but you were working on. Your super ego. That 8 cube. one Jesus said was the log In the guys. Eye. Be working with there's this thing you've been working with there for quite some time now.

Dominic: Looks like you've cleared that mostly quite successfully up to a very big degree anywhere. And others look at you I can see that you're glowing it's all those colours but I'm watching for this one is the silver one like I said.

Dominic: Two things we're going to achieve today is talking we will talk more about finances and how the laws of chemistry and the laws of physics and the laws were motions and the laws of mental events on different dimensions are different things how we can use. What you know already. Connections you keep you with your business. If we have added extra time to go to once you've done dealing with anything else that you want to talk about concerning your recovery your healing and other aspects. So I asked you to check in as you will tell me also you give isn't evidence about. The 8 cube while you were sued but you go that we were working with the last man to talk to you. We were dealing with the superego.

Dominic: And I taught you how to match qualities and you were matching quarter opposites. Or the qualities from the judgements you'd ever made. Those that came back to you. And so I told you about how to play cards and how to play like a 10 how to match forecasts. And how to turn the cards that weren't next to you or those that belong to you.

Dominic: And then I brought in energies with any extra sacred geometry I worked with that to help you with. Finding a better balance and having you understand and clear some things. And you have a big strong interest in the nature as you used to work as a priest as you probably are on the path as a monarch or something along that line. Peruvian tradition was I know that there that resonates with you very much about the rawness of the beauty of life so that's why. So in different systems and different traditions in the world they're different healers and different people who act as such. The college admission that every six one and every fifth one every first one every third one or whenever it is. So at least be worth it with some kind of fairly good cool and are of some kind in their tradition something. Along that line.

Frederica: We said we wanted to talk to the session. We wanted to talk about. Few things now that you're feeling physically about is recovered. Do you feel like you are sitting on two feet you have to be. Starved.

Dominic: I know how it feels to come down. OK well maybe I can consider something to do for you with my training to help you with inflammation. If you visit with when you did the wound cleansing and things. Noticed that they put on extra. You can ask Eric to make for you and look at and see what he sees.

Dominic: They put the what they call they're called the cold pack where they put an extra dressings the change the colour and the extra thing on the ethereal layer almost physical layer of the. Body in their places so the compresses on you with wound care. Has to deal with whatever it's doing. Well what's happened with. You had the butterfly. Things that tell me the rough history since I last saw you. You. Don't find any place studies. No.

Frederica: On the Equinox. I am shocked. I saw. This. One. Then. Something. On this one then you are used to be really close. And I had this experience where. A trained by car. But. This experience around our my life I wish it was me who had all that yes shoes. You know it was me feeling all this that it was like huge shadow stuff with. Rose on that day. So I spent the entire chapter doing clearing you know. Ma'am. Not only. You. And. Really doing deep work and I thought. Wow this is not how I meant to.

Frederica: Spend money. What do you do if you were born. You collect eagle now. Their time of awareness is what it is in the unconscious that needs to come out. That's one of those. There is. No such. Not even conscious dreaming. But just either not just seeing space sense. Being present. How can I take the next step right now. Can be kind to my self. That's really important. So I would take. Every step the first time I went to the beach.

Frederica: I'm a newly built up my strength. They are. You know. Their sheer determination got me there. It was rough. Getting. Up. There. You know. Really took me a half an hour to get back on the beach. But you know it's just a random thing. Q. I'm surrendering to. What. I am I announcing some clients right now. Was there a man. Who is now me just being present. In a moment and allowing my son to. Have that experience and my stuff.

Frederica: One of the things you wrote in your note. Is we're going to help me through my. BLOCK about. Money. Acquiring. Well. Which. Is.

Dominic: One blessing. Is to associate with people with judo or taking martial art forms like karate. But you can pick it up in certain temples or certain religious institutions around the world. And basically you will find that. Basically you will find that it deals with finance. Sixty two percent of the Lords of Finance. The rest is done with. Silver and green (passive income). Red (chemistry)

Dominic: And just like their chemistry laws if you write like you on Wall Street you open up an up in the ledger. It's like making a poem on the mental level or making an emotional. Decision for somebody it affects the morale everywhere. You could open up and you could write log numbers the right hand of numbers down. The right sequence. This is a result of the calculus course you did with me overnight.

Dominic: Where you learnt how to use these fractal's which is a tool who would listen who is hopping with energy the Taurus which gives you physical energy and stamina. But it's also useful for love like you give them parables are good books to finance books talk about how to make us happy and the wealth and finances have light all these spirals all these. Numbers at the bottom of the pages and that extra-large numbers of people looking at the moment you can mentally clean they can see Inside the physical copy. You also get another kind of second book Inside the Book or three books in a book. Where people write these for like especially with the self-help as in the body as in psychology section I saw a book recently by a psychologist at the back. You most likely you don't need this book. You physically happy very stable financially you have this and that you see the vehicle all these levels you learned this book it all is for other people is stuck.

Dominic: I said Oh thank you Miss Lady. No no. She's a clinical psychologist wrote a book on money-making. Because for me when I'm a life therapist I have to deal with people. Who deal with people to deal with their work relationships their family their money issues what job they need to do what job they need to move into in life to me but it suits them more their energy their depression their much toward the middle their physical healing really. It's a very challenging very all-encompassing job so I've learnt some things along the way because you're already a Money Masters coach and you've had lots of people.

Dominic: If you quote the right figure and ride circles the right business in the right figure for people it's like an absolute figures at one point five five one point five five. And you want to push a certain product. I might be able to tell you all about that. Well I know the laws of chemistry. The basic laws of physics for treating relationships of people's dreams and medical issues in motion pushing all these things. These can work for money as well. Money works on the 11th dimension.

Dominic: So I'm going to try to give you and I think I have at least five sessions with a bit of advice that I could give you for your. Money. BLOCK. I want to do something for you because you've been faithfully clearing your. Super Ego. The next thing you might want to clear after this was I gave you remember when I gave you the holders from the 12 down to one and I said pick the number and you said.

Dominic: The next thing might be to work on the Taurus which gives a physical by eternity. Because it energized you more. When I look at the 9th dimension you've got a male Toros. I can give you energy and intuition energy and marketing energy communication energy ethics energy. Morality and energy. And one other thing quality that I forget there's one manifestation. You might need more communication might use marketing you might be energy and intuition. Like at one point I gave you energy in intuition fatherlands accredited in you able to do things and can see them do very much that you were energized and even the sleep was affected. So it was a lot of energy to be. Learned how to handle. But I gave that to you and I just changed your torus a little bit.

If you start working on the Taurus and you about 80 percent of you don't do it take a two or three months to finish the rest of my guidance. It will take you two or three months maybe only going take you a while longer. But if I give you the energy in marketing or anything because I think you see what it is your business the best and it is kind of see what it's about for you your business there's that the next you look good from that perspective if you want to. Look at it from a different opinion and see what no. Maybe it's just changing my now but the book that you book is full. I've been to your website before and look at everything. And you're already a money mentor and you really have a lovely title. On your. Title Which is very lovely because it's balance between the masculine when the feminine.

You must be ready to be able to deal with everything you have assets your businesses and your history with Mike. Of history finances and if you have seen America really in poverty. If one of these this did or not obviously it means different things really. There's been a lot of activity recently on the financial run to make money for people who have contributed wealthy contribution to society in general people who've been left with incomes and you know lots of things have been kind of working a way to make sure that had happens the people who have worked for the Earth have been cleared or they fall into debt. There's how much they're giving back to society. There was there's an activity going on was before it crashed. And crashed but it was that was hard because of donations from business leaders in the world.

A couple of weeks ago and the stock market should have crashed and. They said they must make sure that when it gets to people whether they're less than for them it must be based on how much they've given back to society. Now when there's a law where there's a legal thing it's like somehow it's limited to things I want to tell you when you have the mind of a person but you might not.

Dominic: If you have any skills from any degrees that you've done. Like a psychology degree. It allows you to look at people and diagnose and to hear what's going on in their head and all that certain levels of it 10 meters away to 10 miles. So doctors degrees allow you to the like that you need from not just problems with blood that you need to upgrade on to see what's going on just to see where you're going before you cut. And that kind of thing anything run through a degree if the technology that you're only school your high school certificate will agree isn't there. I mean if you legally and you don't have any of those skills that you don't fall in the extra sense is it anything that you've got from any of your training any along the way. There's a link to that for the way immediately. Said I think the same thing with laws. If you've had bad money trouble and you black this everyday person posting bag or you about credit rating with this thing there are lots of things that affect finances. But it works there is not other banking does make new money it can't fix no atoms.

There's a fixed number of emotions one person can store and deal with. There's a fixed number of money that goes around different people so now the longer you think about big businesses like maybe said such as a company like the about like a scientific company like CERN presses the button basses atoms apart if they don't get the right one the press won it too much. Joy took so much energy and then they have a lot of countries in debt. Laws can change. Like for my healing like I've noticed like in Washington. People in Washington I can heal a scar. Minutes. Here 25 percent Colorado 50 percent. The only thing a change is my bit of dizziness and my ability to change it is the mobility through law to act. According to my skills. In different places. Parts of the world I know that my feelings are and my skills and I use my cancer and all the things that I do. I can use a hundred percent of whatever I use in Britain in certain countries in the world. This is how log light changes from place to place.

That it's got to do a lot of the laws and people who are looking for the grey Halo to become. Self realized. We'll deal with all the laws they've had in their lives or whatever they believe they've had. Let's talk about all the stuff that they've had up till now. We have those in their contracted areas with them to he'll make money or do things another had to supply back to themselves or the like or the energy and all the information or the knowledge that is a time found in the past and balanced about anything you've had for free. You have to pay back.

OK. So. I will need to see some figures and test your psyche and I need to see certain things before I'm able to help you with your business. But I could see at least five sessions worth of time to help you completely. You have really mastered more things about money than most people. You only need five words saying you need like maybe five more lessons with me for militants practical things. Show me your new balance for me how you do the new log life. Show me how you do the new balance. But to do that to do them correctly you need to know what information that I need from you. And then I need to know once I do that I can know how to adjust red Torus which is not fully clear to accent it more towards marketing communication.

Intuition what roughly does it need extra to change it for you to change things just a little bit. On the other levels to make sure that you can be. Free to do things and you need some you. The exact nature you want your money your money is something that people make money. You say you have a school which you were you were people apprenticed to you and you told them healing. Now that you had your Saturn return. You think you were doing anything different from any.

Match to. Be where it is now. That's. How. I came. Into a new venture as a married couple because that means in a masculine and feminine part and then the company can run really well.

And when it goes through in your mind we can go through all the different things. Fact is that I'm going to show you. What you know already. If you use the right log light and Mike's been given some lessons on this and I get a dream- mood module- me a while back. To enrol in like. You know manifestation a country like a business of any where to do business. And you did of course a while ago with me you did that with them with you know with them. I'm not clear and conscious or hyper You see it. He's studied over the course and I said. I would just register him for the right stuff. Because these last five years I've done I've been big being more like. Attendant was it saying. It became an image later. I'm not a market leader or anything.

So they were having different stuff and not for one cycle in this day some of you were very aware of that it was you can see them during the day time is now you have much more vision than you had before you also had lots of things to rely on. Now the one thing that we've developed with sessions with you is your vision. And that's come up really come across very strongly as you've been able to have these very powerful experiences like you said. This is different these levels of reality. Like I said some people who want to call back their drinks and you can look at your dreams that you want to see what you've done before you can do. Side of you'll see a note from the bank from the bank or from the Senate and then the people who are in charge then you dream that the state has had said so many times it says so what do you actually end up getting is maybe three or six or seven hours of. Time. That is your stuff that you can use for your own dreams. And if you're clear on that level the dream that well you can easily see the mental and emotional You can clearly see the pictures in your mind's eye I can clearly see what you're going to get. You can actually preview a dream before you go to sleep.

Okay so. If you can close your eyes for me and really try to do some of the stuff visually with you.

So the best thing I think is to be aware of certain scenes that we have to be aware of in conditions in your world and aware of the world. I'll let you know. The world bank debts countries where you are in the stead once your current conditions are like it's like if you're going to be a chemist and you were going to mix the water and then torture you like like in chemistry to. Might well have observed certain protons. Protons of certain molecules by locating in some laboratories in their buildings. Now I'd like to make to have you meet different kinds of states in chemistry. You need to know your temperature

OK. So basically the following things that you need know you need to know. Exchange rate. Repeatedly with two countries. People in Europe ever deal with two figures. If you are dealing with the Euro and then there's a local currency that they have. To deal with the dollar you've got to deal with how well with the rep with the with the bank (omit) in America which is the Bank of America. OK. So basically if you work with Bank of America. You have equal rate. It's one dollars one dollars not in folders not working at any different rates. That's one value that you need already.

Then based on the average income earning the minimum wage an average income earned in your state and in the world that you're advertising a product to its local It's international. You need to see and make sure that you your price is going to be correct even if it is. It's like it's been at the end of the day. I recommend people I give them the standing for their business and they change the prices and they say wait. Eight cents or eight 1. In one sense or whatever they need to and they figure to make good balance with their budget balance with their books and then work with the financial.

Pressure on them from the outside. So some of the things that are going to come through to you on the screen. Basic log light that you can have for free. This is in the Bible and it's in Proverbs and it is Proverbs 2 which deals with. It's a story about don't be the rich man and the poor man. The story of the public to the story of the women and put many ways advice financial advice in this one. That's all it can mostly give to anybody. So I started it off with that advice. Basically if you read the Bible there are still tons of tells you a lot of things. It gives you kind of model that you can use to balance your books.

And you live according like I do make my money work according to my past you look at income per second and you'll pass close to puzzles. Your heartbeat. Is your income. That can be based your heartbeat. If it's going to be a good reason to be truly said. Thing is worth according to you and you Mike and yours that you say get rich of the two of you. This is your stuff and your whole you have to work it on by stats. And then there's that running out of money. There's as old as it is it's running out of money.

The rate at which you exert effort will you do cardiovascular work and then it changes your figure. This is sort of like the distance you can sprint and really when you get 200 meters you are not running and not doing a cross-country and your not doing you just doing just what it can get out of it. There's certain things that you can draw which will make sense and you can relate to financial figures. But the first thing you need to know because you have a lot of directions. Except for South. And East. In the compass and we talk about the compass. Mike will talk to you if you talk to him afterwards.

You're talking about what did he get out of his fine natural mood module course that he did with me overnight. Basically when you're missing that many directions. Normally if you miss just one direction. You're going to for but by getting everything out of you who look at all the other figures you'll get. What you're missing. That's followed by the book computing with just how we work ourselves with our properties as well as in this 3D form. So when you see the visual you see your access you've got one access for yourself personally or a direction for yourself and your for the east west south north. Intuition Where you headed investors relations with. Products import export and then how you can do this according to a pie chart.

And then you've got also something that connects it to the earth. Connects you to your. Big direction Compass of the Connect to the Connect to the grand connects to. The laws of economics that are in place everywhere. And how you relate this compass how it relates internationally. Before you have an international business you need to know that your alleged product your pricing. Needs to make it needs to make sense.

Currently though I see that there's a there's a shadow hanging over where you have it in full. Do you see that as a composite. Brown. You see this image. And you've got you've got everything in West. Everything in North nothing in South nothing in East but some of the bearings in between covered in the desert shadow having over.

West immediately you from information I can tell that you somehow on a credit card or you a mortgage or you're not a bad debt or you have. Blacklisted but this something along those lines something like it's really forgotten about an investment or something like paying back all university something along those lines there's something covering west. To me when I look at this I say there's something blocking some backlog from so you might be from here it might be that might be it might be what you've inherited from somebody else and they had a battle there of luck and money.

Which caused you to investigate. Who is it from and what do they do. And then we can see how can we start to resolve it. But there's a certain there's kind of ugly business operating with out west which is largely to do with. Equitable income. As it an equity S is liabilities and threats. East is expenditure and then north is your future is your business its future 100 percent with the times and what you need to do to adapt. That's roughly how it comes down to. What do you think this energy or this with this dark energy is how the West could be. [crosstalk]

Frederica: With my doctor. My parents . When I was younger.

Has anything along the lines of what I've said is that any remotely true.

Frederica: I don't owe anything. In bankruptcy I knew at that condition anything I've done in the past to be open to.

Dominic: The directions energy that's targeted so it's not working effectively for you. And you worked efficiently with money before you made lots and lots and lots. The only reason is it means restricting it to this like some of those more taxation more into properties that you pay tax on all what's. What exactly a study loan could be anything. Could it be you own one. One balance of one dollar estate and keep. Your credit rating is pretty high.

Frederica: I must say its almost 800. For. Time. You just gives me a rough idea.

Dominic: How dark does the darkness look it's on the west side of that compass. Now do you happen to learn that the cloud isn't as bad as I said it is. If it's not as dark it's not as bad as when I said. [crosstalk]

Frederica: That it's my sister. I think that I can.

Dominic: Say is you got to just check for me what is what is in the black mist around the west and then you will know. What you need to resolve to use. It doesn't matter what the service is and no matter how much they love it with a decided purchase is between self. How they think their intuition and they'll be guided. If you have a clear access to income like that that's kind of blocking if we can fill out. Why does this darkness. The darkness is as follows.

Dominic: So the pictures will come to mind. Much if that is true for your body you will feel and you see into your eyes that you will see what. Roughly is holding your business back. These are coordinates on the earth and is hiding place to the world of the dead was born on. Let's relates to everything these the business is the function of your business. Our hub in full of fire while acquiring others quickly but certain functions of calling on in the air has moved. Are you one of the abilities of the gifts of my business.

Frederica: If anything you know they are words of way and which years ago I wouldn't even and their billing for my services I would be busy working for my services and I wouldn't offer my services. And. So that was out of balance. But it was more around my own self worth. Death deficit in my self-worth.

Dominic: OK. The wonderful thing about this really tell me about business opportunities with your husband the money thing is that he has west and north. And East. But not enough to get the other direction to make a business dealing with him. That's always going to be much stronger than doing it on your own. The south is representing in. Liabilities and Straights threats of he said as equals and as equity minus liabilities. Liabilities and expenses.

Dominic: Have little to do with how you manage the wealth that you have that you have income how you spend. Do you spend it wisely according to intuition or are you biased. In which case you don't know the direction filled in you will use your ego and you will use things to tell you what to do with it rather than what is really important and then you your biased or by reason.

Dominic: Don't worry. They seem to be slipping away. Okay so you've got your eyes open that's fine.

Dominic: You think you can. You've got you you've got a rough idea. I would like you to close your eyes and look and see what comes up next. Free association.

Frederica: Yes situation was the. Picture like balls circles happening and then kind of dots hoppng in the middle of the concentric circle. So you look at me investigate the myths. This is where we find it.

Really. All it does not your paragraph doesn't do current. OK. With any really good some historical issues to deal with. You've got some overlaps and they've got names and numbers of people and figures and things like that. Attached to it.

Frederica: Yeah there was one also there was something when I was 14. My. Birth. Anyway I was taking money out of my mom's purse. Why.

Dominic: Ha. Does that figure in there. Is there that amount within the date and the date.

Dominic: So I would need to say if you're going to be going to be making peace with income to me it's your business and whatever business you do. You need to somehow investigate these issues and write them down and be that explosion during therapy with each other.

I could explore anything therapy with you and see how do you forgive yourself. And how does your mum forgive you and how can you kind of make amends for what you did that. That's holding you back a little bit. There's not one big things like bankruptcy well being black are many small things from many years and years and years and years and years ago.

It might not even be from this life it might be for inherited from some it might be. Many things you inherited from your parents or from the people you've inherited from. We see all these minuses Venn diagrams. Then you see these little charts with all these people and their figures and everything else. Whatever events come up.

Frederica: It just has the feeling of being old. So spare me or nothing. Like. This breaks my heart. You know this much. You recognize any other dead sort of people the imams. But you. Also. Like Thomas right. Some minds think Paul was changed you know. And you're looking at you like you know how it's so happy. You act. The opposite of thought. And then just read it.

You know it's like that. For happy. That's how that's my boobs. Who. Old news. Someone yourself your family or someone that you inherited from. Obviously. Got a lot of.

Well you inherit certainly by the way you can and had land. Somebody's going to rent motions it can be balanced. Sometimes they like you enough they'll give you the good stuff. Then they're you know balanced emotionally in your you deserve your happy. Sometimes you inherit their debt you heard lots of things from people. It sounds like this it was like somebody in the family or you lose somebody you had heard from was. Basically we could see a few of us perspective you want to see. Yourself in a past life or have you want to see it. Is somehow someone there was guilty of. That. It's a big pretty big amount and it's in. Pounds that correct it's a pretty big hefty bill is pretty it's a lot and it's a long time ago.

Dominic: The debt is like a faraway ago. And that was he went to jail for fraud. Where would be today will be called fraud. Stealing money. For his well actually bit it bigger more inclusive definition than there fraud is actually saying. Stealing is not it is accidentally stealing something. Fraud is with malice aforethought. Do people harm. On certain levels. To gain from somebody who's. Financial poverty.

Frederica: Consciously. But it's quite different and it has to be overt in balance.

Dominic: To be candid saying this is fraud. Otherwise it would be called money laundering or many other things before it's called for. Fraud is over over like a million rebellion. He had fraudster you know you know. It's a different story than saying money laundering or to just having debt. That's different.

Frederica: Yes. That is correct. It is a conscious intent to harm. Absolutely because it's gone. It's almost like it is manslaughter and murder. It's a bit different.

Dominic: And then like three or four more points that are not mentioned that are that are that are worth mentioning when it comes to fraud. Some of it has to be answered in a going to. Organization has. Do you have to have somebody with a whole organization before you can say that you are guilty of fraud. There's always people who are on top of a pyramid scheme and this kind of thing. There's the kind of candidates are likely to have all the right criteria.

What else do you actually feel you see in your head or any other indication in your body about any other history items that you are able to clear and handle. You because of consciously or subconsciously only give you things that are able to handle and clear. There's going to sound strange but I'm actually going to counsel you as if these things had happened to you personally. So this stuff in your mind you worked out very quickly. You said.

Frederica: I forgave him I gave it back in therapy and I couldn't charge for back then and then you said that's fine. And then that was clear and you could see that. Why did that. Did you see that happen. So now is there's another market is this mark on this guy who's a chain round his leg was a fraudster. At that time it could have been his whatever that is two and fifty. Thousand pounds would have been close to that century would have been close to billions and really is a little bit like stealing is stealing getting the key to the king's throne room was something you know the treasure chest and just taking it with them was something. She.

Frederica: Something about a girl comes up. I agree with you.

Something like a boat.

Dominic: Who did you inherit from the east or who did you inherit money from. Came over seas from the ship it could be like somebody from ancestors from England or the continent. It doesn't seem like something did you did it it's like it's something that you've got somebody you can heard from. Who ever heard of somebody from somebody that was a big relative. Do you know about much.

Dominic: Let's work with this thing about the boat and the guy who went to jail and he do his time when he's dead so it can't be too much dead left. My guess is that if you have to. And this is the weirdest thing if he had drub a certain amount of money. And to make a donation to a certain charity. For certain amount. It was outstanding is selling figure. I think. At least there's two things that you can do. He did not pay. He paid his. You paid for his crime by doing jail time. But he did not pay for his last meal said something like.

Well a record cost roughly a dollar two dollars something like that. You spend those you live to make a donation of two dollars. To. A certain charity the right charity or the right kind of organization. It's like the World Food Program or something like that because he was he was eating he got money or. Cooking the Books for farmers or something. It is like a barrel this is ancient is a long time ago it was an ancient long long long long long long long time. Just before they get to the new country into the sea and there was to be murder he was looking at agriculture as the quake these paid you dollars to the World Food Program and then give them a donation of two dollars for the Sahel. Or more if you want if you want to but you have to make two dollars. Roughly That's equivalent of what is left and dead and then there will be a rest.

Dominic: OK. Good. So we are now moving through your youth your issue that you had with your sister would like to tell me more about that. That's coming up as the next thing. We're about 10 minutes left so I'd like to talk about that as well and then give you your work that you do for the next week. I see that you're 50 percent of the way through with your super ego. And what you what you've done before so that means that you were. You've got like 30 percent.

That is you point five percent enough to clear in total. In another month or two of doing that and then you can start working the Torus more energy because then it will be if you feel they really do hit the ground running we'll be able to hit the physical right until they just start a new business venture with Mike towards the end of the year and when that happens it's a good thing you're going to punish him as he fixes the directions you don't have. That's that's a really like a huge thing for me says great idea of you and hardly anybody union makes sure that your business partner hasn't what you don't have.

Frederica: Okay good. So.

Dominic: So basically.

Dominic: Like I said this wherever you go to do with Mike is to be a success and might not be different but it would be something for him for his retirement when he changes with menopause something for you to make. Everybody is going to be different maybe education could be any industry. We don't know yet. Well do you have an idea of what you going to be doing with him.

Frederica: Well we want to start doing it.

Dominic: So you know I think it's I mean education is a big It says kids take coaching education printing sufficient textbooks handbooks for whatever it is that kind of thing would be very big yes. He did make money with that kind of business education business. I mean like what about that when the register in the academy and then doing that, that way because then you could then you could talk or your colleagues and people like making a healing training book on healing or something. Whatever words along those lines and I did said you had some point last year you had to like. Write down a textbook or a book or something like that of knowledge that was practical and works for people and all of us things that didn't meet the criteria roughly of what you were doing. So you see the streak in spider is a direct expression during the time when you were recovering is story on to write something down to type on the computer. He's one do many things. That's great. You've really got the Download now. Something along that line will be very successful will both of you.

Dominic: And. I know in a very gifted artists people wouldn't do it because he wanted all the tracks and all the parts of the medical anatomy of a person covered of his psyche the States the psyche. So the states with the conscious physician's decision cycle and not he states of the psyche. Of the slave of somebodies consciousness and this and the scales that it works on. You could you. I know you're a good guy who does lots of work. He's a very gifted artist.

Dominic: Okay so we've covered some of these things and then put you on the show with your sister could you speak about that very quickly and expand what really went on so I can. Console you are correct on that.

Frederica: OK. You know like I have some extra money. When we were settling some stuff with my mom. Just has. Sounds like I. Know a lot more. Well actually she did very well. So when it came rare very little. And so I have some money that I didn't tell her about. Several weeks. On most things and I just kept that still below. Shadow peace with my stuff knowing that I did that.

Without discussing it with. You now. Why. Because. It's. We the reason that I signed up my really. That I had earned. It. She gets nasty sometimes. And discussions might not bring it. I just didn't even bring it up. So. That was my. Year. Going into. Place. Yes.

So you want to resolve it sounds weird but for the finances to work you could resolve your relationship with your sister. Because there's a bond you have between you and your sister in a history otherwise it wouldn't be in the same family. You chose you chose this hurdle to learn how to manage this. This is how relationships tied into money. It's a big thing. Marketing. Tracking people with things. It's it's all about that. So easing when you make a bond with somebody when you make a bond it takes energy. When you break a bond. Makes it when you really break a bond and chemistry. Makes you really make a bond it takes energy the same too.

If you. Release a business you release clients. You make energy more will come. Hang on to existing clients and you try your business please and more and then you don't make that too many every turnover is a different knows it's a whole different ballgame. So along the way in the next five sessions will help you with this. Your stories and your judgements will always be talking about the money and what you can do to improve your financial or the things to really to improve that are easily able to handle. The place. With the fall of an hour we'll talk about what roughly do you need to do so here you need to do something with this that you need resolve with his sister. There's a big figure left on that on that market says Susan. I see.

250 dollars. You see that when you do see anything else. Yes. Yes. Well I figured you'd see them.

As far as I know that's what you should pay her. He was up on his money level from that particular date. At that time it would have been much more you might have to say that type of dollars was a lot in today's terms that would be maybe a bit different. But if you could do all those things before next week I recommend you on some more things you can do. Yes. Why the world. Will pause again and then the previous one was about. Yes. Alone they took some money you took from your mom.

Dominic: You don't worry about that because you resolve as you inherited from your mom. You heard from. That means you inherited less. Does it seem that you took to your doctor or the executive deductible. Therefore at some point. You got that all out. Or can you just say I'm sorry and forgive yourself about it. And then it just went. By the way. Anything that has really been paid for you really done it really and all you did is forgive the person and think about it. And the intangible for forgiveness for this thing if you say because of some of those that Mark has just wiped it out. Immediately. So you only need to worry about the ones that stay longer. Obviously have a certain thing that we need to figure out.

Dominic: OK good. Some more tips and more things from me. More money tips to get you online and who Pampers. With direction and planning. And everything else which is why you are needing from Mike. You have a lot of that just we worth a lot of. But. You've been eating south and west and he's good and that and you get a lot of directions that you visit. So that's really great. OK. Let's listen for a first financial religion. You could be just working on that level. And if you continue working on opposites. I will send some support this week including this session will be one. Small little. Iso octahedron just sit on the side of that 8 cube and then have to act out. I was replaced but left more clear. In the next week. [crosstalk]

I would like you to finish the is-octahedron and then look at the qualities that you have in the cards and see what are the opposites emotionally right down to the geometry for that. Write down the words for them. One time after he'd run. I will teach you how to make one feel so that he can continue to make one in your life. Next more for yourself. To clean up just a bit. You go through something like this geometry tricks I'll teach you all the levels that all of geometries above. Fear them out. Granted that's it for today. Would you like to see me next week same time or in two weeks time.

10th dimension – Muse

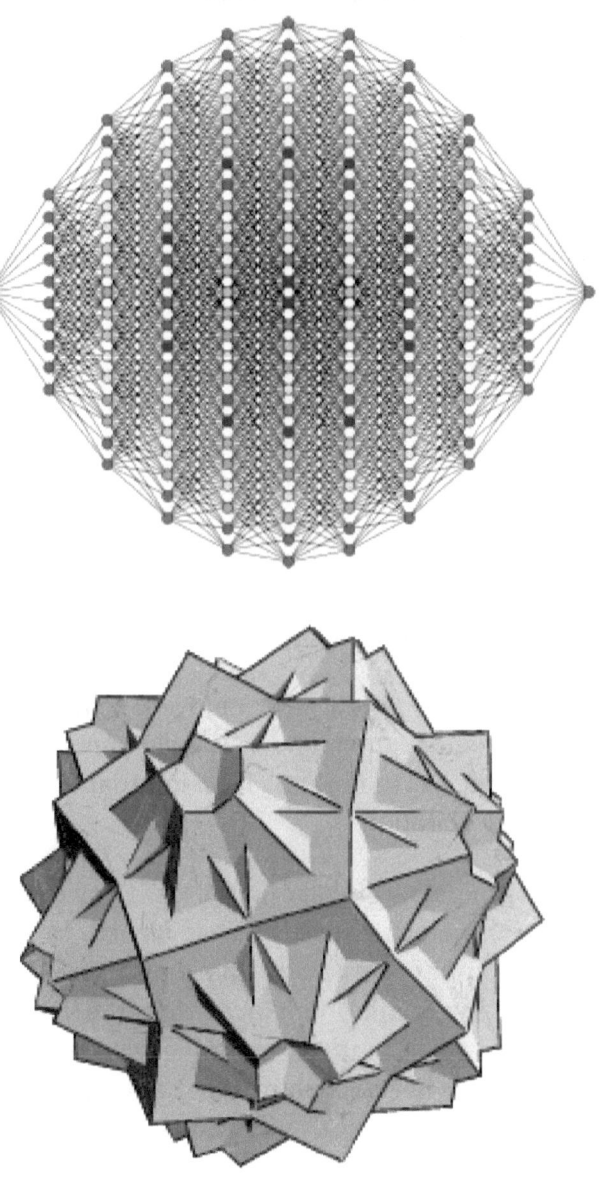

Illustration 5: 10-D (10-gon)

10 dimension E.g. Switching on Musical listening

In late 2014, early 2015 Katja approached me looking for healing in her primary family and relationships. During our time together she would learn how to build and break Chakras. Groove listening, Read life books and glimpse into the future. Very early on in our time together it became apparent that despite enjoying our time together. That Katja, just wasn't overcoming her hurdles. [crosstalk]

Katja was, I would later learn, overly visual and she often would sigh. And I would ask is anything the matter. After a few more attempts suddenly Katja responded with a whisper – I want to hear the angels sing. So we planed this following fix for her cockle shells and relinquished relived and she could see text and receive music.

[I treat Katja with the Blue and Silver Reiki rays.]

It wasn't long after that tat she enjoyed going to a concert featuring her granddaughter and was fully super actualised by the end of the concert. Today Katja is an international channel for world hope and inspiration.

In the mechanism below circular light breaks down debris below dimension 0. Lack lustre love inflates debris from the infra dimensions and caries it a loft to the femto scale then gravity love (force) brings it down. Doom love brings the indicated debris full circle and inflates it in the chemical level where the persons subconscious mind decides abut the back log and delay in listening. The last rhombus 'treble love' reminds us that this is just for listening on the treble.

9th dimension – Torus

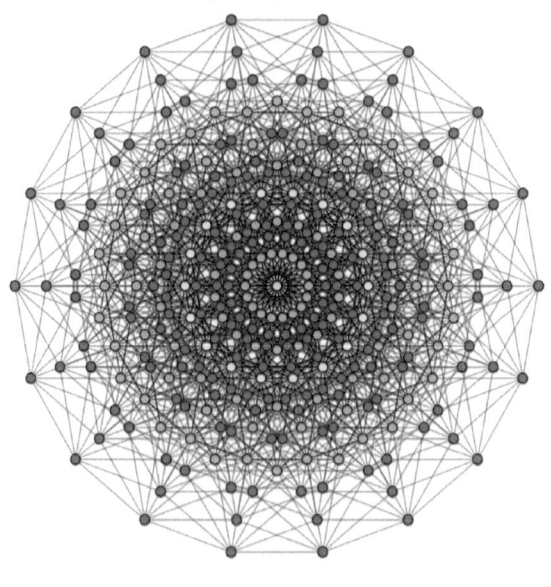

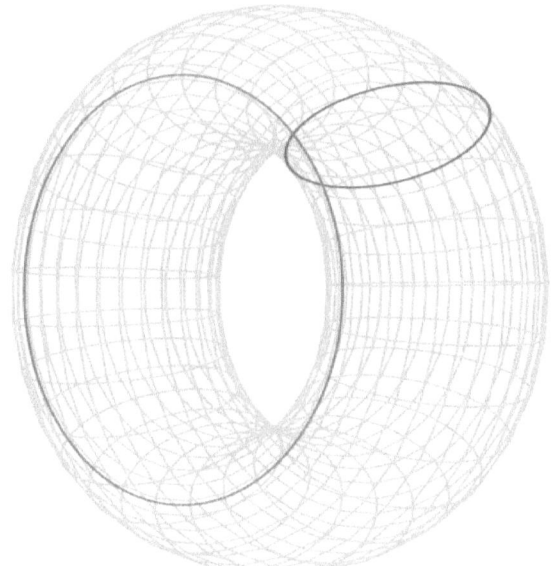

Illustration 6: Single torus (9-gon)

9th dimensional Math puzzles

Torus funnels describe 9 ways in which a person may manifest change in their lives. There are a variety of ways one can clear torus. One is by feeling all its manifestations on our lives (called the path of renewal) a 9 week process. An other is by processing math equations following a 9 week pilgrimage (The path of renewal). [crosstalk]

The path of following ones unique talent at mathematics of statistics. To treat conditions of the torus I teach the heart of the clients a 6 month Chaos math course. That has useful applications in maintaining their wellness (The path of supplication). [crosstalk]

One route out for the former path can be gathered roughly from Einstein's equation for mater and energy: $e = m.c2$.

1. Ethics

2. Equalisation

3. Abundance (Matter)

4. Completion

5. Duality

6. Causality

7. Matter wellness

8. Obstructionism

9. Whole-ism

[One can treat the torus with a mix of Green, Silver and White Reiki rays.]

8th dimension – Super ego

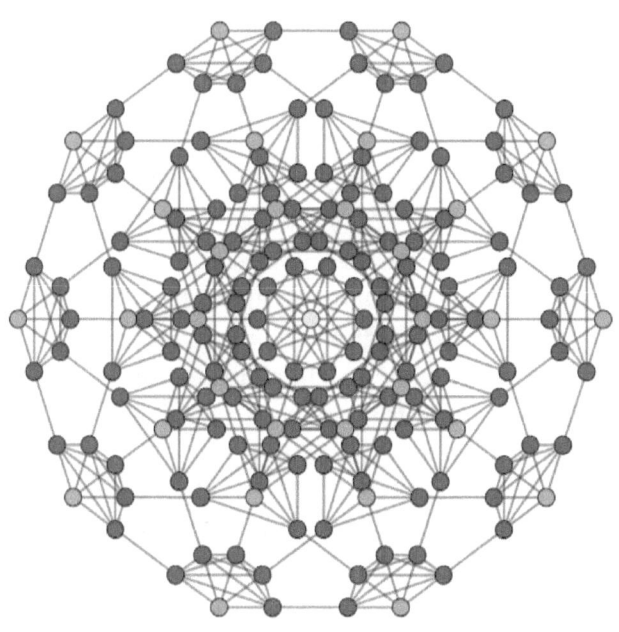

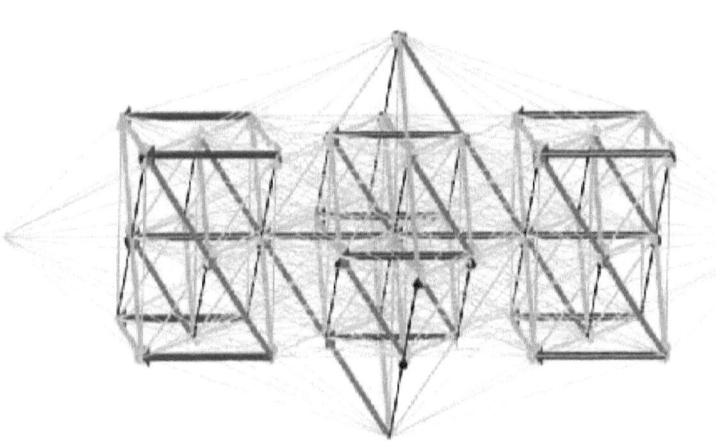

Illustration 7: Super ego (8-gon)

8 cube cleansing

[I treat the super ego with the Silver Reiki ray.]

No that we have seen the basic geometries involved in fixing the psyche elements. Let us try an example. E.g.

A young man and his mother phone my Skype line in a tizz. The Young man called John complains of battling at school with feeling at times inadequate in lessons or and sometimes surprises himself. He complains of rude feelings judging himself during break times. We diagnose a super ego disorder arising in child hood. We talk a while to check interactions. Next I make a metal note of the Geometry (octahedron) used to treat the Super ego – its conformation and correction:

When I first contacted John I spoke to him and took him to place where he would feel safe to speak about what was bothering him at school. I showed him pictures from various fiction series and ending up with a motif of a dragon. I asked him if the dragon was scared. John replied that the Dragon was worried about the school principal. Being called to the Principals office, Many do not make it the same afterwords.

While speaking to John I realise that he had developed DID, was deeply depressed and his belief system of a Dragon council kept himself kept them safe and the magic inside. I showed John a pack of opposite cards. I treated him by applying bio-electro photonics: Of a silver iguana (Appendix 1) - a therapy technique from the African Cosmology. [crosstalk] John's mother, Katelynn, related to me about the divorce from her husband John's father. Katelynn was manic and deeply high on speed. Dry on sleep.

At that time I met my first clinician, American mental health first defence Nancy S. Gold burn. A super fairy being of love and light (yes, its impressive what some people do to degrade the sacred). It would take me quite some time to find my balance in Denver, Colorado my first income group outside my home country South Africa.

Katelynn's daytime personality had gotten the most votes so far from her family and professionals she'd seen in the past. Her dream personae went through a treatment door for speed and was advised on a diet with plenty of potassium. At this time each case of DID used to take me 1-6 months and I had to refer to my supervisor often.

I reviewed with her a scene of dolphins and whales meant to keep her dream state peaceful. I later learned that this was a ritual to send a client into the ocean in the sea of great clinicians out there. I was pleased to find the Clinician assigned to me was wellness expert Sybil Bartlett. Glory trained being of love and light. [crosstalk]

She befriended me and explained all the legalities of operating with international clients. She took thing s to anew level for me and put me on the map. Before this I knew only how to communicate with borderland and customs officials and skipping log life via airmail rather than over water and on a thin continuous loop on some subatomic level yet undiscovered by Science. [crosstalk]

In the case above the young man received appropriate treatment and due to the complex nature of the alters flips too. He competed the Survivor to Thriver course 21 days of rape abuse cases with me. Sybil continued to be of assistance in registering the Yosemite family in receiving the sleep support they needed at a sleep clinic out-of-town. I relieved her stress by passing her psyche through the setting of a washing machine (Appendix 2)

I remember waiting the 7 hours between us to read each afternoon between reports. And applying many geometrics and a pylon for Katelynn. As these client would awaken clearer than the cicrulogi machine and a defibrillator. I believed as my positivist teacher Sibyl says that human connection is necessary to call the errant soul or self back to their body.

I would later learn how what-is immortal in a person depends where they are born on the map but despite the background tradition we all have to clear the ominiverse before finding completion in an after life(s) of our choice. Exactly 47 day's later the Yosemite family that was in a schism was gleaned back together. [crosstalk]

7th dimension – Nadi / Pain body

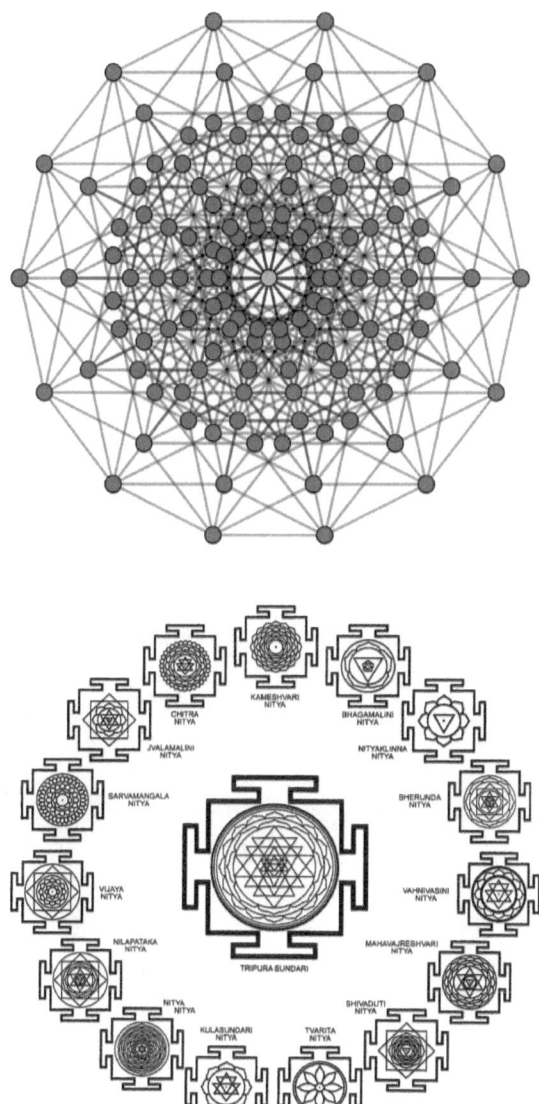

Illustration 8: Gates of Quan Yin - 7D pain body nadis

7th Dimension A glimpse into archetypes

What is notable with this case is that despite the capacity to do An life between Lifetime regression Bridget wanted to stay I n he bubble of the omniverse. Also while the order is not chronological we do see an increase in consciousness (From house 1 seeing just faint while light to 12 fully fledged vision of a tunnel, as an engineer might build through space with fractals and many coloured levels.

This is in the order that each life was revealed to my client (I am using the categories created by Caroline Myss). Merely talking through the houses was enough for my client to become self actualised.

1. House #1: Ego Personality ~ Wise Woman Archetype

    Native American woman. Became squaw of mountain man after massacre in her village. Difficult and denigrating life in mountains. Death, calm and peaceful: merged with light at top of mountain.

2. House #2: Life Values ~ Teacher Archetype

    Old Europe, man wrongly imprisoned. Impassioned to begin teaching spiritual ideas after released.

3. House #3: Self-Expression ~ Servant of the Light Archetype

    Early 1800's, New York City. Kept woman. Pregnant/abortion. Became a nun, later Mother Superior of the order. Worked with prostitutes on the street. Lots of light at death/went up golden spiral staircase.

4. House #4: Home and Family ~ The Priest/Embalmer

Archetype

Ancient Egypt, Nobleman/Priest/Embalmer attending to the embalming ritual and elaborate funeral ceremonies for his beloved father.

5. House #5: Creativity/Good Fortune ~ The Lovers Archetype

(France, 1700'S ?), riding in coach to a party at he fancy mansion of her husband-to-be. Deep, loving connection, planning marriage. Disastrous event (his death?). Died alone, not fearful, but regretful, sad, lonely. Then lifted up by warm, dark embrace and warmth of a golden light bubble.

6. House #6: Occupation/Health ~ Devoted Mother/Herbalist Archetype

Ancient times, Middle East. Abusive husband, 5 children, tiny mud brick home. Loved and protected her children. In older years became a herbalist for the village.

7. House #7: Marriage/Relationships ~ Healing Merchant Archetype

Son of wealthy British merchant family, taking leadership of merchant ship for the first time. Devious attempt of crew to steal cargo. Was shot, then rescued by loyal crew members. Survived fierce storm. Transformed; no longer felt he wanted to be part of his wealthy family. Travelled in Asia. Married. Compassionate outreach to community. Experience of between lives after death.

8. House #8: Other People's Resources ~ Champion of Social Justice

9. House #9: Spirituality ~ Fallen Woman Archetype

England, 1902, poor farming family. Eloped with London nobleman; learned to play harpsichord, to paint, other languages. Many friends, very happy. Pregnant, two children. Very resentful of her children/didn't want to give up "fun life". Strife with husband/affair with another man. Husband divorced. Contracted consumption.

Worked in clothing factory, angry outburst, pushed over sewing machine. Dying on the street, rescued by kind woman. Life review: took responsibility for her choices/felt gratitude for woman who rescued her. Dying, went into white light, big white dome, star-burst of gold light, far away, then surrounding me

10. House #10: Highest Potential ~ Beloved Musician/Wife Archetype

Japan, 1902. Wealthy, talented young woman, gifted in Tea Ceremony, music, art and calligraphy, married handsome young man. Very happy marriage. Difficult delivery (had to choose between life and death), time of confusion and resentment of new baby after birth. Husband very loving and supportive/family life improved. Husband died, son gave hope. Lived a long life with her son and grandchildren. Peaceful, playing her instrument and singing. Died surrounded by loving family. Some difficulty surrendering at death because of family holding on to her. Let go and into the light tunnel.

11. House #11: Relationship to the World ~ Weather Healer

Late 16th Century, SE Asia or South Sea Island. Trained by shaman/mentor, officiated at his death. Practising shaman: healed a girl, birthed a baby, helped by human and animal guides, including dolphins. Connecting with spirits of sky and sea to bless the baby. At death, Spirit Guides all around me, I go into fractal black dots on white background/become One with the fractal dots. Merge with brilliant golden sun, with golden spikes coming out. Go into a white, cloudy tunnel with white star at the end.

12. House #12: The Unconscious ~ Shaman of the Golden Sun. 15th Century, area of current SW United States. As a young man

Connect with Underworld in a drumming circle; I am led by Spirit Guide (tall fox-headed guide that walks on back legs) to meet ancestors. Later, the village shaman becomes my mentor, much love between us. Nearly captured by rival tribe; rescued by fox-head Spirit Guide. Pray in cave and meet Spirit Guides in Upper World who say that I will become a great Chief/Shaman. As an elder, many people come to me for wisdom and healing; I channel wisdom from my Spirit Guides/am very connected with the Spirit World, where I travel to different realities and dimensions using coloured shapes and geometries and the facets of a prism.

At death Spirit Guides (and many villagers) surround me. See exploding white light and sacred shapes pointing into a crystalline tunnel, with golden/white light at the end. My Guide appears, dressed in long, white clothing and we step into a huge, bright crystal/glass room with many loving friends and ancestors; sun comes through like rainbows.

Walk through a tunnel of sacred geometry through my knowing and my learning, hen emerge into a golden sun/light, which bathes and purifies me. In a huge crystal amphitheatre, I enjoy having no body, see many souls, all different colors/vibrations surrounding me in a vibrating cube that is part of us. Many geometries, seem 3-D, but made of energy.

We communicate and are connected by the geometries. I try out different ones. My Guide takes me up via a beam of light into the Sun, where I am being given knowledge of who I am/what I do. The beams are vibrating me and helping me know and understand everything, understand my capacity and that I am a Light Being. School Room of the Sun: teaching me what I already know (but forgot), teaching me from DNA, connecting me with my Original Encoding from Source. Very intense/see pulsating energetic forms.

What is notable with this case is that despite the capacity to do An life between Lifetime regression Bridget wanted to stay I n he bubble of the omniverse. Also while the order is not chronological we do see an increase in consciousness (From house 1 seeing just faint while light to 12 fully fledged vision of a tunnel, as an engineer might build through space with fractals and many coloured levels.

[The Nadis can also be relinquished relieved by working on the client with the Yellow and Silver Reiki rays.]

Parallel present regression

In early 2016 after super actualising a client, Bridget Frost, by regressing her through 12 lives as guided by Caroline Myss archetypes. Even as the end of therapy, Bridget loathed the tunnel and numerous times told me that she didn't want t to disappear in the after live. She had however show rare bravery in journeying with all 5 senses in the extreme past. In looking at her data while the lives weren't always chronological and some lives overlapped despite our best fact checking. I already knew the answer. But was pleased to find my email in-box open one day with a message from Mrs Frost:

> " ... I am attaching, in the order that they were revealed, all the regressions that we have done (so far) for my Archetype Wheel. I am using the categories introduced in Sacred Contracts by Caroline Myss.
>
> I was interested to discover that the two most recent regressions revealed lives in the year 1902. It will be interesting to find out more about this (parallel realities?). ..." bringing PhD's from the west to Tibet to hold world peace. I with quite a racy pace was able to translate albeit only in Navajo. With the three of us.

Shortly after this We logged 3 present regressions with Mrs Frost. The present pulse regression allowed her to break our of her shell. One shows a warm reception with the Reincarnated Dalai Lama born as a new child and growing up a year per visit. Lastly we experienced him as he was choosing his spiritual gifts mine, mine, mine. [crosstalk]

6th dimension – Mood

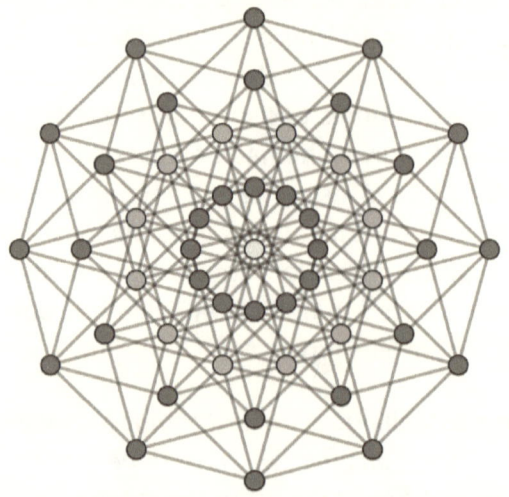

Illustration 9: The well of Dreams

© pinterest Chakwa ala

A triple DID integration session or maybe not

In late 2014 Paula contacted me to assist her in healing and integrating her what she called her soul shards. After a number of sessions we met and talked with Carlos, Olivia and Bea. While another professional might diagnose her as suffering from Multiple Personality disorder.

What was real for her was the concept of separate selves each carrying their shards. Integrating Paula took 6 months to seal the languages and past memories. Through the process of present pulse regression, Paula gained Navajo and Greek by interacting with her doppelgängers. I made use of a pylon to link the neurological, psychotherapeutic log light and watched her mood by applying the White Reiki ray of helaing.

Paula: Olivia has been sharing some memories with me and doing a lot of clearing and releasing listing of different things. The last three. Like almost every day.

Dominic: She seems to be out of her depression and she's ready to unify.

Paula: Olivia. She is ready. I think so too. [crosstalk]

Paula: I think she is grounding through me last night that it was time to start doing.

Dominic: That was with you could feel good when you haven't had a bad night. Do you check in with Louis to see what's happening.

Paula: This morning and I think it is because she was sort of grounding me. I don't know if it was really high. I got a headache.

Dominic: We talked to her what did she say. You know and I just asked if she had gotten it me I didn't know about her. OK good. So it sounds like. Did you. Did she say anything about. Wanting to integrate in any particular way. As the powers that be decided to be. Or has she said anything about that happening in a certain way.

Dominic: I thought we were going with the theme of the tree.

Sense that you've shared all the memories of your home and your whole. Tree of your journey.

Dominic: lovely. Good. Well. It's great news that you seem to be in a state where I would I would I watch you I understand and your sense of consciousness. I. That you are in a state of. Sublimity tending towards goals toward Epsilon from delta which basically means that you're going towards the buddhic level of existence.

Dominic: To that state. And so the sleep mind you and I don't get the best sleep tonight. And over the next the next basically two hours or so you might see hyper vigilance you might experience bliss and rapture and many different kinds of expressions and you know that Olivia is really intending to merge since incidents ensued and sound before the end of the day. You might start feeling that familiar feeling of when he feels soft in the head to have a migraine but it's like meagreness a painful heart. It's cold and kind of. It's like a basically like cold me going to the next and so like a cold compress does that make sense.

When you merged with Carlos you had this astounding experience where you. Had sex with Carlos merged with you know you had basically. He met certain requirements that you ESP is you vision a certain way with things that separate sides of America. But you have this where you didn't express where you felt really truly well come and just be a part of him you felt that moment of reunion. You remember that event that was parlous when he first greeted you and then you started to remember his side of the story what happened to him and you remember Can you explain what happened to you and you went through your life to see what had happened.

Dominic: Drag all your lives that you that now from the moment that you had split from him apart from him. So you said in the beginning of this phone call you said you could just feel the. Vibes more or you could you could hear more of the tape. Could you just elaborate just a bit about that. And stop in a minute. You said that you could when at the beginning of this phone call you said to me that you could feel or you could hear. What was on the treble of the other speaker. Of his computer you could hear that he could hear some sound. Is that right. Was it like music or was it more like.

Music or did he hear anything different. Nothing like that. OK. Just checking. After we said good morning we were silent for a while both of us we just looked at each other. You made a comment. I couldn't quite hear what it was. I just have to go back on the tape and see.

Paula: OK. OK no problem.

So lovely. Let's continue. Oh I don't think there is anything that I can see him. I have a cold and you like an anchor.

Dominic: OK. Remember when I told you we're going I told you that you were going to be hearing when you go out you are listening you would be hearing on the radio or on the air conditioner on the treble the car and so on and so on. Just remember what you said to be what it is that she was in charge of doing this thing for some time. That side of you that's out of your ears. I was wondering if that has started to kick in already but we'll see soon with the integration with Olivia what's going to happen. With your gifts. No

Dominic: probably give it to you at times when you can handle them as soon as you can handle the next gradual phase through. Through the low you've already experienced. When you are younger you Grandmother Florence. Must have made case with the judge because there was a lot of legal contracts and deals about your abilities you give to you is paid to the manager throughout your whole life that you have not been able to access all of these things with Constantine since you were a youth.

And so you had to go through a lot of things where you felt that you were you would try to use up too harshly. Are you going through the moments where you had to move back and see what his grandmother Florence doing with her time. And on my desk when I look at your case when I just when I would greet you on the phone I think I pick up a whole docket that says by 56 which is next year for you.

Dominic: See this kind of trouble well whatever is going to be a big day so I'm going to watch you carefully the next few weeks to see how integration to Libya happens. After especially after September we might need to call me again for a while just managed to see how you do with your guests how to switch them on how to switch them off how to manage to figure it all out. And how does it matter. It's first of all the matter whether or not there might be additional of this two additional as to how it's going to work. You're going to have. Certain kinds of vision of the colours mainly obsessive consciousness will be given that vision.

Dominic: For seven for seven days seven days will be in the sun seven days in the shade seven days in the moods and days in the tide.

Dominic: So you have different vision initially every week you have different kinds of events to have the candidates the next week you might have. Visuals you might have math you might see a different way you might see many things. But it's governed under the law Florence made for you. When you were very much younger. OK it looks like you're busy integrating with Olivia now. Over the phone. It's going to take a while. So I'm doing my best when I'm remain silent. I'm either trying to assist the energy or trying to guide everything along. Course it takes two. It will take at least a few hours. To kind of. Feel normal again if we get any movement or anything now don't worry about it. You just feel really kind of feeling that you felt you familiar with now.

Dominic: It's almost like that different kind of focus that you get when it's almost actually able to multi task to a greater extent. But over the next few weeks I watch it carefully and we'll see how much of everything even deal with safely. And we'll make a will make we wonder what does it take it or go out or we can say maybe. Basically see if we can ever see if we can manage.

Dominic: More. Or less and we can we can try to just try to see if we can kind of. Work out of the like what you agree with that you have with me you manage your sleep. And manage your mood and manage your other guests that are going to be working on. After October. And during this next year is going to be particularly challenging because you have the switching on at different times you may have to be dealing with all of them.

Paula: So that's really going to be hearing.

Well you've also got a pretty good sense of feeling and fragrance for smelling. Gifts that should work for. Certain kinds and as do as rocks that should work full. Perceiving the high. Levels of existence.

Dominic: So basically what I would like to do is I would like actually watch you for a period of a month or two. Into the. Six months at least just to see how you're going and if you feel you need to call me you must come and if you feel you to get my attention non verbally or on my hips said all you could get my attention emotionally. If you get to like a fever pitch or if you get too stressed I will obviously reach out to you emotionally a number but when I try to help you calm down sometimes you can be quite depending on how it is because I can read here that. Each of these is governed like by a totem animal.

So you will have a vision. That will come into you and you have a vision but it's almost like seeing if it's coming in from the Spaniard to zoom out and the feeling of free-falling from a plane. It's got a similar kind of feeling to it and your vision is like that for a certain while. And know this is going to work. It's going to be mainly what is physical. And all your other senses will start in a different times also. But the main thing that you're going to be that way different for you are the vision and the listening.

Dominic: Is just the main reason now is the point. The point is that in the beginning you were going to see whatever we can get and whatever edge marks are left because generally. They fade after two months to 20 years of use or 10 years or decades. You get some. You have but basically as your ancestry you have certain kinds of markers on your body your head your cheeks and so on and so forth. On your tongue inside your organs and in different places in your body.

Dominic: You don't seem to have the one that says that will be the one to see things externally but you have a lot of gifts that see things internally. How. See your dad on Can on Can. That. And. That's basically how I see it. But.

Dominic: You might have had you might have had for the very first couple of first years of your life and see the gifts and see the one that looks for three different kinds of gifts. That's on the spot on the forehead of somebody on their head that allow you to kind of see what other people say or a sight or bubbles or ideas bubbles or whatever it is emotions and so on.

Basically. You have to get a grasp of how your visual sense works. It's what your inner vision of at first. And as we take quite awhile as you've had times you've had it's come sort of like a preview experience before where you have you just got your vision for a few seconds and then it was released.

Paula: That makes sense. Right.

Dominic: So like geometry or so a visual of an image I was for a few seconds and you quoted and then suddenly you knew what was and then you have you it was like you're falling out of the sky almost. And then you were rushed back to it and then you not understanding what it was. Does that make sense. That's where it all starts.

You the first time with geometrics and colours and shapes and after that as you get into it as you kind of deepen your knowledge we will work with our relationship you try to help you. And gradually get used to all these gifts. And as you do this. You you basically get into different parts of the melody also with colours. You it with its native American people to help you understand and you get to obviously some of them have Navajo origins or links or something on it or some few things with Semitic writing on your cheeks. And that there's going to be. Some. Because there's going to be some.

Paula: Need to be to maybe.

Dominic: At least find somebody in the Mediterranean or rabbi or somebody to kind of officiate certain things at certain times but then many have to do with listening. And listening to music. Tones. At the moment all you hear is just this slight high-pitched. Whistling kind of sound. That's what that's how it all starts with visions listening and he gradually builds on after that so we can be very clever we can try to manage your case and see how we do over the next six months.

And then maybe you need six months with me to figure out how it all works. And after that you can actually see the world and you'll be fine. After a I would want to make sure that I still maintain you sleeping cycle for a while. It takes at least two weeks of adjustment to make it just a cycle super cycle two months. You'd imagine that after all this integration it must be difficult to kind of orchestrate your sleeping cycle and some nights you wouldn't have had slept. Most nights you should have gotten some rest.

Dominic: Well luckily for you, you had a very good you have not melatonin available need lots of extra minutes and with your you from to help you natural remedies to help you with getting to sleep. And I'll be watchful and obviously if you feel that you need to call me you can call me but I would want to ask you for these next six months to see how you get into all of these things and demagogy through this process. It may not be like magnificent visuals. All of a sudden it's my fault in graduating. OK. But I don't have any idea how it's going to work out the deception is for many people. You know the only exceptions we together. Also there's also when you get there there's also a time to talk about how it happens when you put different is pieced together. You can look at it if the pictures are different and they kind of combine different kinds of those men. They see things differently.

Dominic: As well but not it's a beginning of a lot of fun and a great insight there of excitement and exploration and bliss and it's it'll be a pleasure for me to be your guide along this process like we have been for the last few years. Embodiment of your gifts.

And so with this we're taking on this long reach that you get to eventually get to the point where you do this life has you believing and you're not the soccer mom any-more you're like courage and you're doing holding retreats and you are fit you feel that's just in your body you feel together you feel you so you would be basically at some point we had a glimpse of what could be. And I very much still like that visual for you visit it makes a lot of sense that you could you haven't been through all this experience with yourself you would be a gift cancer to other people. One of the most that I read in Navajo on your face that's actually said cancerous.

You have certain talents. You're a great listener. You will see when you'll see how it changes when you speak even this even if there's a speech on the radio. It's not just few people who switch on in this planet. There are a lot of people it's almost like a subculture that people can figure out who in their 30s most mostly depending on how they leave their home. Depending on their restrictions of their value system their tradition of their family.

Paula: They were farmers really they weren't. All from Kurdish speakers, Luddite or whoever it was.

Dominic: And you have lots of ancestry from the Middle East so you'll get to work in a certain way or maybe in countries in certain provinces in certain cities we might have to we have to be aware to think all of the south. So I'm not going to say anything I'm going to promise anything immediately until we see what's going on but you had some kind of episodes where you can have seen or felt or heard things briefly for a couple of minutes as you had come a preview of what it might be like. So and then on top of that how do you manage to get this hollow about which animals are going to carry you through the decades and through the weeks. And which ones are you persist with you through your lifetime.

In the beginning it's going to be eager to see which meaning you will only hear what is carried to you in physical sound physically of a long distance and that's how it begins because otherwise you will get too absorbed in into many other things you might be able to follow you might not be able to follow your daily life. So we need to make sure that you're very well guarded and ready to take on all these gifts stop and running gradually that's how it's going to happen.

And you've said you've had experiences before and you've had and you've seen that they were switched on for a while you saw them for 30 minutes over an hour 15 seconds and then they kind of close down again. So we had a couple of different things and you reported to me over the last few months what is what you've been seeing. So this is kind of what are you preparing for when as soon as you get to the point where Olivia merges with you. Nothing is lost.

Dominic: Olivia's talked you about the bad times you won't be able to read them as much any more you probably will stop thinking about the past. Obviously over the last few years thinking about the past that you've had your soul shards and you parallel selves you've been thinking about lots of time geological time span and then historical time that you call in your soul you've been spending time doing things that you have a much more future focused. You we focus on the present. It's really going to change the style of your life to a certain extent so have you ready for all that. All the exciting new excitement as you said is that you've been so hard and suffered such a long time.

I can't wait for you to really just receive the joy of just doing some things with the people who other people just get in different countries or different systems that is get rid of the bad and they have this virtual neck suddenly seeing all these things doing all these extra things. And it's occupation so you both have a life and you certainly deserve what you want are you going to be receiving. These perks from that spirit. It's it's going to work again. I fear less fearful about causing harm Yes. We're very happy. I think that's still kind of there. Thing done. Life is so restricted because you know I felt like I had to do.

Dominic: If you can find it you see kind of in your mind's eye now you might just close your eyes and tell me you can see some colours. To check which you see where we.

Paula: I'm seeing something.

Paula: Kind of a reddish tone a red color. You have kind of greeting there on me the darker color of my hair. The first is the kind of thing that is just the computer. The first break. After flying out the side.

Dominic: Being how can you see that it's written it's right across your whole minds right. Can you see that.

Paula: Maybe the word color kind of saw orange light is kind of I don't even know how to describe it that it's not really. So it does seem to be like the phone in any favourite cloud or something came the way there's more I like over here the different almost the yellow color and it goes like maybe a bit different more on the size.

Paula: This is just my window because it was greater for me than this thing. Right change is a change is going to kind of come out and it is good going into. Greys with many punches and then grey dark grey. Remember Gray was in many circles curves like different and dark grey some kind of colour. [crosstalk[

Paula: Now. There were a few there with that were more than a spot of white light and black and bigger than just Points in my head sort of like. Just the little.

Dominic: That's lovely. That's very good land beyond. Can you see the grey. Which images then the so that was different. Speckless of colour. Can you see the Gray underneath it?

Paula: Not sure a flash of a Charcoal colour.

Dominic: And it came and went and we just trying to focus on the grey. If we could focus on the burgundy the red kind of red burgundy kind of would be fine just to focus in one color at one time. So now we just need to watch and wait. I've got if you can just you can you can keep your eyes close and just come in what can you see as it changes every of the next couple minutes.

Dominic: This is D-day. [crosstalk]

Paula: there is a bit of yellow gold. Change it's almost a green colour. Is like a black or grey.

Paula: Service is cracking the square. I read that on the different levels of functioning with various types of objects. And what they are is different.

Dominic: I either have more dimensions in them that move angles or they have moved sides or they have less. It's like exploring now exploring your subconscious. And it is different angles. So you might see just that geometry to be more flat and have not much. Around it. That's one way of appreciating the different levels or different levels of dimensions that we've been talking about a lot.

Dominic: And it's really interesting because it means that you have pretty good vision for colours and you're going down and not keeping you. I'm keeping you to this space with your team and my team we're trying to just keep you safe and steady in a certain kind of zone for a while for at least the next two hours. So it won't be too exciting but at some point you could do meditation. Maybe in a day or so that you do meditation and then see if you can see the colours with me next week and we can see from. Around from red to. Back to Brown to pink to green to white. To gold silver and so on and so forth right away.

Dominic: I am pretty good at meditation. You just didn't have the kind of guy to tell you what stage you were in when you were meditating. Now you have some appreciation for palate which is switching on which is really wonderful. So you are aligning yourself just nice this much which is great as you are show me that. When you walked in at the legal level which is a charcoal grey kind of. You kiboshed certain rules and you said that you love yourself. And we said I love myself. Some of the other way. The religious rules you build around Florence rules on top of it her rules by herself. And they fell a way. I don't know if you felt anything at that moment. I was. Trying to figure out how to balance you out so I wasn't focusing completely on everything I started I've seen the on trying to keep in a certain state but I would not I would have to ask you later on but that hark back when you were one of the chains that come off.

Paula: I've been feeling Energy. You know I don't know. I'm not sure. OK. You don't want to.

Dominic:Yeah. OK. So you can you see the red again. Are you seeing the charcoal color.

Paula: Every bit of blue in the know blue colour I can. Think of things that are kind of fleeting.

Dominic: Are you trying to.

Paula: just to see you know they keep seeing it just keeps sort of morphing and moving.

Paula: It's exciting.

Dominic: Turned into something interesting.

Dominic: Delving into the dense. Extensive consciousness in the subconscious it's really interesting. You know how does when I read your books and New Age books and the books and they say that you know you have to be a page or chapter looking at random looking at all the colours and you go through it. And this is how the vision starts warming up. And you know how it is and then use that body just before it doesn't get connected because you just didn't see stuff inside your head of colours at all. So you know now you starting on this journey and it's a great thing.

Paula: Yeah its fun. [Paula was released as a robin 1/5 way through the integration]

Dominic: Obviously when this starts to work really well is time to think on different levels of different intuition that you might come up with kinds of intuition to solve problems in your life from a different angle. And pick up a bit you know pick up a big kind of like a bubble of sand to see what's going on. And it's really good. It's going to be really it's going to be a wonderful journey of discovery and it's going to be. Sad. Here we go.

Dominic: Do you have any pressure on you. Ears. Is it take me or ring at all.

Paula: Oh. Yeah.

Paula: The ringing now I hear another.

Paula: Static. Study. Either way you know the key is and this is the kind of thing.

 Paula: Who. Now is static right.

Dominic: Like white noise.

 Paula: Now. And they're ringing. OK. This.

Dominic: How you feeling about your parents? These days.

 These days. Not recently and you said they are they send more regards to Regards and you kind of. I don't the movies you've been on this because as we've been talking about the family tree. Time you spent in your gene pool with your band. Do you feel more on it and do you feel kind of more. At home in your body and you feel more. Supported by you or others.

 Paula: More things that I had to release that were related to some of my green from my grandmother and more support and maybe I do a CV or. No. I mean when he said that I would like to say sort of just feel. Different. My dad passed away I guess which is probably good because he's here more and you're going to find. Her through a lot with that relationship. And my mom the feeling that I don't know this seems like there's still something there with my mom. I don't know what it is but you've been great to work with cancer to figure out what's going on with the sisters. And since you don't use you you're reaching to say whatever it is that you're trying to gain.

 Paula: So much has given them power over them so it is a sign that I am in this life and you know I gave them so much power over me. And I hope what's come up I never really know you know comes up and then I'm asking for it release released listed and hoping that I don't really know. You know like there is so many layers of things for I just turned over so much who I was kind of let them sort of decide things for me. OK good.

Dominic: I would like to know who. Students to do with your forehead and he's just between the left side or the right side.

Paula: Yeah you know I'm going to have a headache today and I think I am on the side. I feel better than I did. That was for all of the. It In. The rain. And. There. Is waves of light waves of colour going from the right towards the middle over green maybe not often not only greenish colour now one is a burgundy one at a certain Its a certain level it could come on in the kitchen.

Many just toxic responses where your reflexes are just doubled in crease. And things look a bit like two-dimensional. To a point. But you right. But I'm glad you got on with this and you have to with awareness see the colours and the neat that is the cube underneath the red not just the charcoal and that part of the red but in between and in between the six states there are four states basically. You see yellow congenial kind of blue phases and then the silver-grey with lights on her purple pink and green. And then they have different kinds of steps the kinds of colours and you extra you describe them really well.

Dominic: Paula you're going to start understanding how these things work and you'll be reading on the jury you can talk about different colour protections what that means. Different people. The knowledge is not always pure and it's not scientific and it's not always accurate. And I might say the Redbud's quotation with you root chakra too but also things is not really that it's really the best said about GABA that's what I that's how I see it. At this time. It's often used by people in the martial arts. To respond faster. But it's not it's not generally possible to stand in for longer than 45 minutes.

Dominic: I think is pretty sure that the best source would be from me over the phone. And after that I'll give you some books and these you can read about. Manual with it's unique like this was in the book. By a man who was in it this time just for his whole life. The people who live in states of dreams in different levels and detain people without having heart bypasses and different levels. And then the things that happened during during surgery I could have passed that to you. And then there's also the. Other books that I could find.

The talk about the levels where he's like Barbara Anne Brennan great astrophysicist. Since then she has this she has a very good idea about healing and multi-dimensional levels that you might be ready for and it was time. But through research and through application you will find intuitively you will find out what does it do for you what it means for you rather than what the books say is you are more of a deist you have your own ideas about your. New journey you don't always want to buy the books. And remember there's always six sides to every hair. And there were many traditions that would fit everything that there not everybody about one or two days and they're all correct.

Dominic: But that's only from one perspective all the time.

Paula: Where is more important what it you mean not what anybody saying it is. It resonates with me.

Dominic: It resonates with the people of it. Teaching doesn't resonate you just discard it. And that's how you do go through life. And I'm sure you folks need more energy and besides you might feel a bit lonely not talking to anybody any-more. That I'll be here. And only a phone call away.

Paula: I write I'm very. Ready. Not to have parallel selves. Yes. Knowing what is. Going to be really nice. I've still getting soul shards. It's been getting so short. So far so hard to retrieve yesterday and still don't really know how many more goals I need to go and get. It they're like you know I've cleared some. Current green answer no one ever and then after they're clear to release them I can go get those shirts.

Dominic: Scintillating clear. Writing the very transparent. And you seem to be very free. Kind of social of your own you be doing this for 10 years. And you might have had a few more lost time with once used and degraded out to be any more. Started getting negative. Images and I hear what they are. They are quite right you know and they have. To do with different things.

Dominic:You know. Something I have no idea. Really quick they're just sort of far. From being on that are you seeing it move you see move she is going to work if you just. It's been over an hour but I just want to see that you are progressing really well. Would like to just monitor things over the next. That you are in the right state to make sure that integration can see perfectly well in the beginning. I thought I saw that it said two hours couple minutes. And I understood that it would take about that much time to integrate.

Dominic: Be careful the rest of the day and just be mindful of the process. And when you close your eyes you can just report and say what you see. And it's going to be OK and we'll have to have another talk next week to see what's going on. I don't mind what you need for another. 45 minutes just to see what's going on. Could you see what could have been a plan. Negative. We know the man is dark and it was like a light into the light grey you're right we were something that. We knew better than to be learned is something that you know to be. Because it was sort of that negative. So I don't. Know if there was some doubt he wrote that.

And so my advice is not to connect to try to just keep you being so living for such a long time with the wife white daughter. In dreams and connecting. To others is to live deep in and to approach the low-level as much as he can anywhere between the great trouble burgundy red would be fine. Keep your focus for at least an hour. And if I've got time I'll ring you again just to see how you doing.

Paula: Namaste

Paula, a mother of two, now enjoys being the link between all her parallel selves through the journey she undertook with me of bringing them up, healing the fissures and applying common sense.

5th dimension – Mind

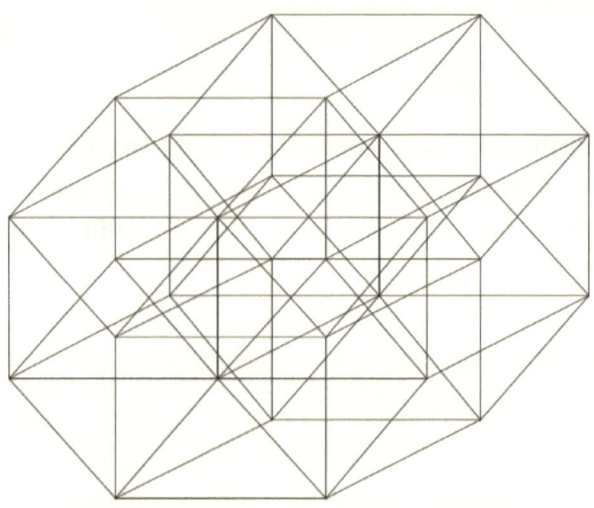

Drawing 1: Debris of the mind Artist: Eliza Fitzgerald

Case study for Hypnotherapy for substance abuse and advanced knowledge of the passing of a second mentor

Constance contacted me again in late January 2015 after becoming ½ self actualised wanting to explore her over-eating, alcoholism and relationship with her daughter. [crosstalk] While applying the talking cure I use the Green Reiki ray to relinquish relived her spiritual debt.

Constance: Hello

Dominic: good morning to you.

Constance: Good morning to you. How are you doing.

Dominic: Really well thanks.

Constance. I'm happy to hear that.

Dominic: Great. So the last time we talked which was last time you did about a series of about four or five sessions with me. Well we worked with issues about your family of origin to your mom and your dad and he went on emotionally and then worked in past lifetimes. We got very emotionally and you got you very clean emotionally and mentally so we were able to work with. What looking at past lifetimes and futures and things like that. And we worked really well. And since then you have obviously continued to do really well. But you've written to me recently and you've said that you want to work on relationship with your daughter.

Constance: Yes.

Dominic: Pattern's you have with her how you react when she reacts with anger and bitterness or whatever it is and you try to pacify her and then it just has a lot to do in that instance. And also then you want to work on. Your weight possibly. And then the drinking is that right.

Constance: I really did. You know if the drinking sounds like it's you know I don't have alcohol every day and I'm fine. Yeah. So it's not a huge issue but it's so unlike me in general.

Dominic: That makes sense it does understand that you're not there you know the kind of person that you don't have some troubles or any of these things. That might lead one to pursue drinking. But it is not an exact match and it's not good it's just the occasional glass but it adds up. You don't want to turn into that want condition into becoming a big problem.

Constance: Yes. Exactly.

Dominic: OK. You're familiar with the process of the tapping and the. EFT and the TRE and you're familiar with some Sacred Geometry that we used in the past to help you with some feelings.

Constance: Yes. [crosstalk]

Dominic: I will work on the emotional cause of why you might be drinking. We will work on past lifetimes indicated with you and your daughter to why she's acting the way she does. It was just her nature is not going to solve every issue and we'll look at ways you can alternate it's just so different options of how to interact with her.

Constance: Yes. Great. And also with my weight I am exercising more than I have ever in my life and enjoying it. And that is just it just seems like it's more than just it's not a physical thing although it's a stick.

Constance: And I already eat really well. [crosstalk]

Dominic: I eat like a lot of you very healthy. Obviously you have not had the intake form that you sent me and all of those things. And it probably has an emotional onset to some way.

Constance: Yes I agree.

Dominic: And remember you had you had like that energy when you've been asked what you think the last very long session we were on. You have an energy of the tummy. Of heaviness around your stomach remember.

Dominic: Yes. And then we cleared it and then we thought OK this is obviously just like.

Dominic: That could have been foreshadowing a bit.

Constance; Yes be interesting for you to kind of check in and see what it all means.

Dominic: Wonderful It has a place to be working with you again because I found you really pleasant to work with and very agreeable and very very very willing to work and feel your issues some people say that they're not. I find it very rewarding and very refreshing when somebody is comes to me as if I want to work on this. And they do and they make progress very quickly like we did the last time.

Constance: Yes. Thank you. I am. Yes. I'm thrilled that we're reconnecting and I'm ready for like another layer of you see.

Dominic: There might be a mental thing like conditioning to a little bit of we believe humans getting emotional causes first and then working on the mental level and then on looking at other things to see when you look at it it's a historical issue from lifetimes before we can look at doing that process as well.

Constance: Great. [crosstalk]

Dominic: then once you clear on most of your levels you know individually we can work on some family things if you think anything else might be impacting you. You can do better. We can talk about anything else and then we might even do something with transgenerational transference although you have enough I think with. What you got from your parents. It's up to you and we might if you feel the need to work on ancestral patterns and things if there is something we might be called to work on that as well. But we got to start with emotions and reprogramming and we'll see how we go from there.

Constance: That would be great. I mean my family and I are such a. Vague world. You know I could spend the rest of my life working. Thanks. You know what I mean. You know. I'm open to doing whatever I need to do. And. You know it's. It's. It's such a large.

Constance: Place that I you know probably never get around it or maybe that's all

Dominic: I've done a lineage process that takes a person a month. They look at 14 generations their family they see what they are doing. They get down to seeing a meeting with them again and a bit or having contact with them spiritually. And how they work through it by. And if you're a visual or you have listening it's a very nice very lovely journey at that time when you talk about like that. To listen to them just to have them to guide you to say what's happening. What did they offer what kind of values they had.

What kind of journeys to make bank through the values of the family to what happened present day. What were the difficulties what were the problems and to really all of that and so this is what I had from my from my ancestors and then you can see the Soul Journey is another aspect we can look at as well separately to that what happened to my sojourn in some of the people again. What happened with the last one. And so on and so forth. Very glad to have you back and if you can you would like to get started so if you would close your eyes for me.

Dominic: We've got a couple of issues to hit and I'd like to get started on I'm trying to get access to what is going on and why. It might take us back to a past life. It might take us back to a moment in this life. May take us back to just a feeling. That might be from many sources. As you close your eyes and as you meditate give me permission to lift your energy a bit to.

Dominic: The crown you should see white light. And the frame should come down in future sessions when we get. A little clearer than this. Emotionally mentally and you can sustain yourself for more than an hour in the course. Instead we might be able to go into a point where we you could project into a six dimensional light body. And. Walk around the library move you know look at books look at look at your life other people's lives. And futures dreams during that you've had a nightmare trying to guide you what's going on subconscious. And lifting it into you know the conscious mind. To work on. For the time being I'll be trying to display images and things and might just go back. And it be blurry. Might be back just to quickly if you could grab an image of what happens to have on something. That is what we developed and working to get to. Last time I was doing work with you I would just send images down to you in the way that I'm aware.

OK. Do you see the white space in your mind's eye?

Constance: I do. OK great to see you go into your frame. Circular frame. I do. OK great. So I'm going to try to get access to the root the cause of I'm going to give you three things when I try to reach the root of the cause of the. Right again. Eating.

Dominic: The drinking and then we will look at the relationship with your daughter as well. I feel it's important to get to the best. To be as neutral in this impeccable and kind of clearest possible. Centred and clear as possible personally. Before we start working on yourself and other people in the family and then we can work on some historical stuff from all the other families or groups or the world or whatever else might be affecting and impacting on your. Angst. On the screen. As a mom and from a from a past life was a moment like this we got to that moment that six lifetimes that come out as linked to the eating it seems also seems linked to abandonment.

Dominic: This is Calcutta India that you are you obviously as you can think about that relate to your present life about abandonment. And then obviously independence leaving the family at 18 or before then around that time and just escaping. Well you did. But this goes back to an issue about abandonment. And it's not all the time about when you've had poverty in past lifetimes where live in poverty and this is this lifetime. Can you see the picture of. Where you are this young Indian woman.

On the street. You were either orphaned or abandoned. Ran out of a home because it was difficult or whatever it was. Then it was abandonment. Comes up very strongly the feeling of abandonment its red. It's angry it's kind of attendances. It wasn't me that is. Just me. And my. Feeling of being unwanted. And there are lots of components involved in this about poverty. Because that makes a kind of sense psychologically. You are eating to try to fulfil this need of being abandoned and you want to have plenty to say that you don't ever get a situation where you have poverty and you're trying to compensate by doing this and it's unconscious of these things these drives for many lives or their six lives in total.

Constance: Yes. Yes I think I struggle with a lot of these feelings of. Abandonment. I wanted it. Yes. I mean even when I do something really wow. I just skip. But you know I can't get over it quickly. Like I did some sort of piece of work with the group. It was actually quite good for a moment. I just felt like oh my gosh that wasn't. Very good. They probably didn't like it you know because that's.

Dominic: A view from that the criticism we can explode as well. But one of the big emotions in your field is this thing about. Being abandoned being unwanted and then. This. Is storing up for the future which is about storing up the tissue in the body gets the message it needs to become bigger. It doesn't want to hold one in poverty and then the next thing is. Probably being critical of yourself. Wondering about if this will be enough.

Dominic: You know I mean that why worry like you said that came up with that feeling that when you've mentioned about. The group that you had that food system wasn't enough to then like it it's probably not. Now that probably we're checking in to see what is happening. Being objective you immediately just said.

Dominic: Oh no that's not enough. It's not it's not it's not right and it's not. What maybe it's not. It's not at all. It's basically this big thing that comes up when you talk about that this feeling of just isn't enough.

Constance: Right. Right. Yeah.

That's the feeling we can get into the story of intellectually why that happens and the conditioning and things like that we can also work on that in some sessions as well. OK. So working on this one I want you to do some tapping. Why are you watching. Indian. I want you to feeding her feelings of being abandoned. And what about the feelings of some depression and there's poverty and there's some other things happening there as well. She took great heart.

She liked to do dancing she liked classical dancing. You know everyone on the street but she did. But in and out of Calcutta supposedly. And she did some work as a dancer. Which was a great. Amount of joy and relief. But what I'd like you to do because the. The opposite of poverty is contentment which is a tan feeling is Obsessive contentment. And then the other opposite of.

Dominic: The opposite of something like being disowned. I think someone is parenting. And the opposite of being abandoned. Is. Something like salvation is like being saved. Being found.

Dominic: So I'm going to try if you give me permission I'd like to try to shift this energy in your emotional body. And I'm going to send you first. This obviously but being found I mean the energy of being found of being of salvation which is a silver emotions and you want to go along with me into the sacred geometry for this one. It is. Oval with two hundred and eighty-six degrees. And if you give me permission to send his energy to you why are you feeling this in this person's life my life will just go back to all the lives that are. Occurring behind you know kind of egging you on the whole time. And we're going to talk about. It is going to feel through the solution is going to feel through the solution for all of this. Are you picking up on these feelings of abandonment. We're not going to talk about it. We're going to try to tap into that entirety and all of these. You can tell.

Dominic: But I first want you to try doing the motions. Do you allow me too?

Constance: Yes I do

Dominic: I can work on salvation. And it's with 45 degrees something about being found and there's that kind of energy of love being formed and that's the opposite. It's an interaction energy to. The energy of being abandoned. I'd love to go out. I'd like to cultivate. When you feel issues of abandonment allow you to cultivate the opposite. The opposite of salvation is being found or being wanted. Acceptance.

[During this moment I download into the clients pons a personal geometric connection sequences. So that she can let me know about any weak moments. I can then adjust the gaseous neuro transmitters.]

Dominic: Acceptance is actually is it is even closer I think to the opposite of abandonment. So. It's it's not pluses between salvation and acceptance and being found there's not really such a great concept in English for. All the emotions we've got. We've got over 500 000 words in English but emotionally there are not that many.

Dominic: Talk about all the emotions and all the different states of all the opposites and have an opposite of everything but that is in the experience in our in our world that we live in. There certainly are ups. You can have your journal open and just write down. Rhombus 45 degrees. Being found and then the feeling of being found. And appreciated and accepted and then it's really about acceptance at the end of the day to feelings of being critical of yourself. Has. Opposite of criticalness is kindness being kind which is yellow with a trapezium to you draw yellow trapezium and so kind to work on the critical and then the Salvation is the sober over 286 degrees.

111

Dominic: And it's about acceptance accepting yourself that what you produce is not. And it's also got to do with being accepted by others. And the abandonment thing is about. Yes the parents of that love and accept me. Not so. So working with those three opposites and those That's the sacred geometry that you need to generate the feeling.

Of to generate the feeling here it goes you're doing the Rhombus 45 degrees. Nicely done. Does that seem to alleviate the feeling of abandonment.

Constance: Yes. That's just great. Yeah.

Dominic: We're working we're trying to hit the emotions with the opposite too. To just recycle you. That's what we do we really care the chakras we just check opposite emotions the whole time a lot and seems to fit the mind. Let me get limiting beliefs and complicated things. We start talking about how can we get a line in our head and we keep hearing a song in our head. We've got a certain word or negative thought or something stuck in our head when we do a certain action or something.

If you're aware of those sorts of things most people are aware of the emotional things most people don't set up emotions in their lives very well at all. But if you care emotionally generally you're aware of the mental content and stop irritating thoughts intrusive thoughts. So you want opposite thought absent mood. And then it cancels it out. And there are very many other techniques and things I can show you to do the same. OK. So that one worked really well.

Dominic: And salvation did a bit to a certain extent because on one of the lifetimes this is the great thing but you just see all the potential lifetimes of that are behind it but there are six main ones. Behind it. So we might as well look at them and when you honour them and you look at the shadow and you look at what

112

you don't want to face or you look at what's causing it. The problem is often alleviated because you give it conscious and engineers say I see you. I validate you. I love my and I see the lesson reflected in my own not a nation I see it today and I can see I can end this vicious circle if I want to but it's a choice.

Dominic: So. You know a lot of things you can do the people work with vision is trying to treat this drinking in and eating disorders and lots of things in mental disorders and things symptomatically like in your states and things that are about that they're treated with person and they don't go to the emotional cause of why the person broke down in the first place or multiple times it was multiple times in their life the traumatic fact is that you have to do therapy and that's at least 50 percent if not it isn't the solution. Which is sad is a lot about just treating the symptoms. And right here we are going back to the causes and began to look at the past lifetimes and then we go into the feeling of FMI and then we try to out we try to just try the total opposite dimension experience of it. And when we do that very well we are free and we liberate ourselves.

Dominic: OK a symbol. We go back to Samoa. Can you see the picture on the screen.

Constance: Not. Quite.

Dominic: Is it now.

Dominic: Here's a different kind of situation here. You were outcast from the tribe because you had Ideas that were revolutionary. That went against tradition. This a lifetime way you had ideas and you had out of the box new ideas about how to believe and what to do and when it gets traditional. You found solace maybe with yourself. So you've written those three shapes and the opposite feeling down into a channel. I want you to just look at this lifetime with Samoa. And sit with this and as it comes up this week these feelings will come out because I'm reading them from the bull. You were going into it and looking at also. The. Nature of the feelings I mean trying to get out of the get out of the chakras. Out of the chakras and then recycle the feeling into something different.

Constance: Because I have this. Very frequent. So in terms of my work my like I do really great work but I. Hesitate to write about it or talk about it too deeply because it feels dangerous somehow it's it makes no. Sense in this life it's. This could very well be from that and then I'm sure it affects many essence of your life. So we just great this next one. I'm going to analyse this emotional profile you can probably feel with the feelings are you can probably tell me even quicker than I can what's going on there. But basically it's abandonment but it's also it's not just that it's more like. Not because it's physically he could deal with that. He was able to live in the wilderness but. It's like.

Censoring that there's this feeling of being censored for insight sense and full brilliance sense and full of genius. Of new ideas. It's it's it was a mental thing. But it's also in that in your mind it's still there somewhere. But first and foremost you must work out the emotional. Part of it as well. Which is basically the thing about being ostracised. We talk so let's talk about the opposite. Be welcome and valued. Being welcome. Feeling of being welcomed. Do you mind if I send that to you.

Constance: Oh thank you. Yes.

Dominic: OK. Here we go. Can you feel that energy.

Constance: Yes. I do.

Dominic: So we can see that you know these lives obviously having an impact on your present life. And it does seem to make

114

sense when you can't find the cause in your own mind. Anywhere. That its going to occur in the past somewhere. And it's interesting how these things work together the abandonment and the being ostracised and being. Is not being welcomed and you just feel that you also you know have sort of more critical for some reason in your mind. Work on. There is an audience in hypnosis when we do working with the mind.

Can you see any pictures on the screen. In your world in your mind's eye. It does help to see it visually or experientially the motion or whatever else is behind it. Because it does mean that. You can then start processing all these feelings and once you process them. That cause of origin or issue is lessended and you find peace Your fear and the right and you just have no reason to want to drink to put a reminder to eat or do anything. Out of if it if it's you know unless it's just what you need to get by. That day. So. I think it's going to become really great if we get through just even if we just do one thing or think about the abandonment today. You can just do that today. I mean you walk down your journal does that sequence of feelings you go through every morning until it's done. And they will analyse that probably some other lifetime. Another couple of angles to it which we'll explore in future sessions.

 OK. Here is another life So you've got that in your journal you need to use. Well I love it. It's a red triangle 35 degrees welcome energy of welcome. Yes and the opposite of being even easy acceptance again is while here so you've gotten a ready. Acceptance and then being fat acceptance is some. Kind of like not a right angle triangle is 120 degrees.

Wonderful. All those all those four or five or six shapes that kind of summon that feeling for you and kind of activate that feeling inside you. And then your next week will work on some still some other things as well and angles that aren't that we need to work on. But we are making some progress here which is really wonderful. As long as you feel the shift is happening you know you're doing great.

Constance: Thank you so much. This is exactly what I need.

Dominic: OK. So let me see. We can look at the story behind the minds me today to think about why this might be happening. And it allows us to say OK gee this is affecting all these different things in me. It's just this whole life from the gun and it's gone it's over. We can't let it take over. Now we have we have to put it to rest. The other one of the things about his ideas and this is I asked him about his ideas. They weren't accepted. So the opposite would be if they applauded or honoured for that. On a mental level it is. It is gold.

The energies were sumoned and it being loaded for that is the opposite. And it is. You might look at more of a mood or aspect in the mind in some way we can probably do some work on that as well. But for many of focus on the emotions and then next week we can start working on the mind and then it gets a bit too boring. The opposite feeling then I'm just going to try it's not that serious while it's being loaded being uploaded and loaded and appreciated for your offering and the opposite feeling here is like this. Is something about exultation too. But we'll get to that aspect. At a later stage. How does that feel.

Constance: Oh I am so I. Just really good I just so really that we're you know you were you know I think we're working. So I'm. Just so. Happy. Really.

Dominic: So these are the ones if you feel these feelings coming up and the point of this therapy is to trigger these emotions they're not suppressed in your body or anywhere else and you chakras that they are. You may feel like a tumble dryer. And as they hit you meant you use these opposites and that if you don't for me beforehand last time. To do your tablet gets a little older and then suddenly whoops it's gone.

So tapping brings it up from a physical body into the emotions. And then this emotional process you work with the feelings to nuke the feeling with its opposite. And until it's done that's the monk standing history of the Brahman and from many lines I see six of these were women to it immediately. So we're going to work with. Another angle to things. That was one thing about being ostracised because a man's life doesn't have physical abandonments this was a mental abandonment. But. Still. He had a bit of a connection and he was still there. OK. So his lifetime.

Is also very interesting a lot of these angles on it. I think it's really interesting. His lifetime. I mean you can see this picture or you see colours you see images and all. It is in Saudi Arabia what would be today Saudi Arabia it is. A time where you were. Shut away at one of the first asylums it's about the 15th. It's the year it's about 1586 and it's one of the first asylums I ever had in the world. To treating mental disorder that had no clue what to do. But it kept people away from just because I can understand that. So you were once again ostracised there because you were hearing voices and you had. Insights all sorts of different things were happening. And they abandoned you and they put you aside from society and there was stigma. Because of a mental issue. Your mind and psyche broken up into different parts. Because of stress and trauma and that like and they just locked you up and they just want to do that put you there and there was no one to treat you with what to do. Really other than stress and make sure that you were. The capture you need so you weren't abandoned like that. But. The family didn't know what to do that and understand. So.

Dominic: Here again it was kind of like. I'm different. I am just. Here. That's just another angle. On this story. So here again. If this has been something that you've struggled with gives me something to just with in past lifetimes it's been it's been a number of them. And if you feel that. If you are ever going to stay with you have a something like a breakdown that I'm here and I've got about a 90 percent success rate. You know I don't think I'm at risk for a breakup. Good. OK. I've learned in my therapy. It's what most of us are conscious at some point nobody is safe completely. But I just I just know that if that's it that's if you get very sensitive when you get to that level. You open up the mind and that is what is going on there. Many people have these things that aren't even noticed. They're there they just cause stress in the body. And emotions and they don't have any source to anything. But you are a very A type personality. I don't think you're addressed in these lives. It was a bit different.

Constance: And it's good to know that you should begin to feel anything like a breakdown that I would request your assistance. Yes. OK.

So here we go. We are looking at this life and he was just. You had some delusions and you had some other things that were notions that weren't quite right. But it wasn't it was again some stuff that wasn't in society is didn't you know you were you wanted everything he wanted independent and do your own thing and all sorts of things. Not as much as in that lifetime in some way you are from the trap of having these new ideas and revelations. So reflecting on his life what kind of feelings are coming out. Sadness.

Constance: Yeah. My guess is.

I'm going to try to alleviate this with some sound healing. I'm not sure if you can you can I'm going to apply some sound healing.

Theoretically spiritually and if you can hear it if you hear the sounds that can let me know if you don't hear the sounds that's also OK but if you feel relief and the light is in you can feel that trapped becoming lighter and clearer. That's And I'm going to proceed with using this. Method of. Integrating and see if it will be a benefit to help relieve the throat and the sadness. The Sorrow and the sadness of sort of this is happiness which is a trapezium. 45 degrees is a happiness love. Let me just transmit the happiness first. Here it is. There we go. Can you feel that suddenly is lifting the sadness and that sorrow.

Constance: Keep. Yes.

Yeah. I mean I think my is. It's communication you communicate a lot. You do a lot of work around that. So I'm just trying to find a way to balance that on the mental level. Just give me a minute.

Dominic: Is that getting better? I'm just about to use some. Sound healing. Modalities at the etheric level close to physical. I don't know if you can hear that at all. Any sounds or at least toning or anything. Else that I. Manufacture and then I use.

Constance: Sticky. Yes.

Dominic: So again when we lifting it up and we getting it we getting it active. We are lifting it up and we trying to kind of move. Get it to move rather than be stopped. And once it is in motion. We can start to work on this process we can start to. Accelerate things by working with statements in the opposite mood. So here we go.

OK. So I want to use the next half hour that we have the next 20 minutes to work with some. Mental reprogramming. If you give me permission I'd like to connect. With some gentle Reiki just to kind of smooth things out just a bit and maybe that will make things feel clearer. And if I, if we want to and if I see there's a

119

necessity to we could start working on reprogramming maybe some thoughts if we need more trust or is just tell me and then we can work we can work on this maybe. Two or three weeks down the line.

Constance: Yes please do.

Dominic: Well I can apply regression therapy again because this is not emotional it's more mental.

But you can focus on the congestion in your throat and you can see it moving now and integrating and changing and shifting. So here it comes. Here it comes from some energy that was specifically designed to work on the mental level. And the thought is I'm guilty of so much. I would like to say you would replace this with. I'm not guilty at all. What do you think is appropriate. I'm not guilty at all.

Dominic: Says that I'm too fat. There's this judgement. Say rather the "the love I have for myself is reflected in my outer appearance". Or some other positive affirmations. Something like that. Whatever it is to look like myself I've made it and. As I connect you can you feel the energy is shifting and can you feel. The thoughts coming up and then subsiding.

Constance: Yeah. That's the sound coming through that is trying to work on the mental level to try to cleanse. It's like just lots of light lots of white noise and it can be like lots of high notes and it's an odour. And then I heard one time she said she heard. Rushing water. And some sing cymbals and some bits and some that and some high notes and all sorts of sounds. And then she felt relief in her body in her mind. And that is something for the emotions. [crosstalk]

And that's it. And she asked me how on earth did you do this is about seven months into water. And I played it from the internet and I found you know this massive of soprano singing all the nice lines and Sanctus and Benedictus is the right notes and I ordered them myself in the right order and I came up with that with the right to the right kind of nature to do that cleanse. On that level it really kind of deeply cleanse. The mind so this is a process we're starting on that's going to last for a couple of days.

So we are putting it on so that it can get out then you can start to just work on things and if there are intrusive thoughts that are kind of. Things happening maybe. You know you start to you'll start to notice that there's less congestion anywhere unless you know your. Thinking process is just more agile and there will be lots of benefits that you'll see that will come to play and I think it's important because there are a lot of beliefs that you have about limiting yourself about the work that you that not know that it is not a people that's not welcome. That it is insufficient or it's not. Great enough or whatever it is. And you've got to you've got a couple of mental things to do where you limit yourself with quite a few behaviours. So. It's going to be my pleasure to help you with all of these things and just give you assistance to the yourself.

Dominic: So are you feeling a bit of resonance and fearing happening with fuel in your throat chakra and then. Some light ringing in your ears.

Constance: Yes I know. I just feel it. Throughout my body and spine: I just. Love them.

So it's correcting some things in me doing a number of things to shift energy and to kind of make it easier for your body to kind of flow into your. Crown Centre where you see the guidance and the pictures like we've seen in previous sessions. And it's just a lousy that kind of comforting to perceive things that you need that is in your mind and your body and your emotions. So that's

121

why we working on the emotion such a lot. In this session because we wanted to make sure you're as clear as possible. OK.

Dominic: So we've got about. 15 15 20 15 20 minutes left. Let's have a look and see. What we can do to spring clean the mind a little bit further. We're going to look at some beliefs and the wonderful thing is we can shift in your mind even if you're not hearing the intrusive thoughts or anything like this. It still affects you because you keep thinking. That something there's something behind. What you're doing some unconscious part that is just tripping you out. So here we go.

Thought I reject you I can't take it any more. I want you to replace this one with something to print. Something like. I can handle something like that I can handle it. because that statement can be used to support many different lines of thinking. OK. Now and that feeling behind it was like no one knows me or I didn't feel appreciated or I don't feel honoured or accepted or something you know and. I feel lost. Say. Rather say. I accept myself something like that. I've said. It. All I've said. OK. I've made it in. Different thought but it was very nice restocks that thought.

OK. And that's about the opposite mood as well so we can find some stories to really kind of get them ingrained and we can kind of want to. Lift them out. We can. Work on the exact opposite nature of the thought need reverse the thought. By saying its opposite or you work with the opposite mood. You work with the opposite mood and then it becomes neutralized no longer. Has a certain tone or has a certain kind of. You know that sort of quality to it is diminished.

There's another one. I feel heavy. I feel heavy. About who might replace that. I feel like. Yes. I feel light-hearted or something like that some it goes another modifier there. That's the basic kind of

122

angle of the thought that made it in the first place that one.

We spent many lives collecting this money and adding things to the stuff and I have a rhyme which is that all sorts of things. It made it into many different levels. For many many lives. I want fulfilment if I can find the right one. Because I think that comes at. The right step made in the absence and made it in. Your mind seems moot. Most of the time is a little bit more luck a little bit more. Free and then I hope that your emotional profile. Has. Changed significantly.

It's it will take time but if you can do your homework during the week and you can just do those opposite feelings and generate them. Dissipate those feelings that will just make the world of difference to your outlook and to the issues you're experiencing in your life.

Constance : Oh wonderful:

Dominic: Is the main topic we're focused on today was the weight gain. And we looked at some of the mental beliefs about. Acceptance and abandonment and whatever that was going to be from looking at the past. Did you manage to see any of the past lifetimes some frames and some images and still things and some blurry images and some feelings and things. OK we got time we still got about 10 minutes so we're going to work on. We're going to work on. One more life. They said there were six. And we've seen three of them so far. There's some more to work with next week. One more this week. Gosh what a lot of. What another angle of things. OK. I'm just glad what's going on here.

You see the picture in your mind. You are in Carthage and you are. A woman. And the society is patriarchal. And you're wanting to be a philosopher and you've many philosophical ideas and he wanted to become a philosopher under student student philosopher under one of the big philosophers in town. But. You managed to do some classes but it is it is so difficult because they don't accept you just because you are. A woman so be not accepting ideas mental things. Not being sane. All the revolutionary ideas and then other things just not be accepted physically just because for whatever reason just not being accepted at all. As a child of my young age the thing about abandonment. And all these things that range along this theme of acceptance.

Can you see that picture some doric columns that you are now in Carthage. I kept trying with someone that came up with some strange thing that you had. Ideas that were anti the gods or something and you had like. You know very good and you were just trying to express your. Own insights and intuition and your own understanding when to get against the grain. So in the first century it was a Phoenician colony. And. This is early in the first century. This is in present day Tunisia. OK. So feelings again from this life what are you picking up.

Constance: Works just like the same stuff.

You mean is that what you're saying. Yeah. Pretty. Rejection. Cost. I mean. We've picked on we are picked six lifetimes that seem to resonate with underlying causes for the eating. And I thought that could be the way to get out of this could be really good. And. There's this the two others. That I hope we can get through today as well it seems to be about acceptance so we work with the acceptance love. Acceptance. Acceptance. Do you allow me to send the energy to you again.

Constance: Yes please.

Dominic: There we go. So then the feeling of contentment and satisfaction because you feel that you've been hurt. A little bit.

Constance: Yeah.

OK. So we've got about five minutes left so we'll try to tackle one more. Live. And you've you've gone through the main ones we need to focus on those feelings that I will continue to done the. Sacred Geometry and you put down a list of the names. Of the opposite qualities. So let's look at one more. Here's something interesting here's a lifetime. OK. Do you see. There is another lifetime that you had on the beach. This is a small little. Island and it is she. Ibeza. And here while you're not totally outcast. You're not totally. Welcome by people. Either.

It seems to be for a more emotional reason this time. Because in this life you are a young man and you seem to be very empathic you seem to pick up on other people's feelings and you can never knew what was wrong with you because you were just changed or he would you would just feel different. So you were prone to drinking to find a way out of your. Situation. Can you feel this about the emotions this young and passion and some. Of just being overwhelmed by the emotions feeling of overwhelmed.

Constance: Yes.

Dominic: So I'd like you to tap this out of your system but it's still buried in there somewhere on. Tap on your on the karate chop point. To an inch or two and a half underneath your pinky on the side of your hand and say even though I hold even though I have this. Even now I was even though I have this overwhelm. I deeply and completely except myself.

Dominic: So. Even though I have this overwhelm I deeply. And completely. Accept. Myself. OK. So. Now to the top of your head like we've been doing and you've done it before in the last sessions you did with the. Top on the top of your head and you say. Oh go. Fontanelle body find on top of your head when and say. I let. It all go. I let it all go.

Dominic: And you feel that shifting. OK. Well that brings us to the

125

intercessions. I'd like to bring you back into your body. Yes. Much lighter and much happier much bolder much. Better for the experience. Would you like to make time to see me next week.

Constance: Yes Lets. How is your Monday .

Constance: Same time. Same let me let me just. Happy. Yep same time would be wonderful. It's an. Important day because it's my first day what a great thing to do on my birthday.

Dominic: Why shouldn't you something we can do something special. While working on ... His idea finishes. Yes wonderful. Wonderful. Oh that's great. Great again to be in contact with you again. All the best this weekend I hope you just practice the safety and just when you pick up these emotions coming up and you feel like you're in a tumble dryer just to work on them with these opposites that you can see how effective they are. And I mean it just neutralizing whatever it might be baggage or whatever it is that you're carrying around.

Constance: Yes. Thank you so much. OK.

4th dimension – Chakras

© Erik A. Blackman

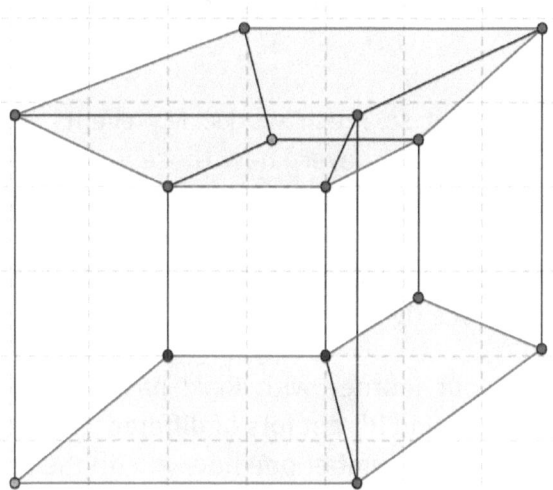

Illustration 10: 4D hypercube -chakra clean up kit

Cookies all gone

[I apply Red, yellow and White Reiki energies in this session about halfway through.]

Dominic: So in today's session you've said that your listening has opened and you told me that the email that you can hear a song sometimes. Sometimes you can hear some words and you can even communicating a little bit with the flowers in nature. And it's really opening up. Tell me more about that.

Lily: Well mostly just seeing that when I'm in the shower it seems like it's about to sing to me. So I know that's nasty. And there is some actual words to that for two days.

Lily: I keep thinking every time I listen to those great times day the staff in the music won't stop. But let's make the time. So now let's go. But it was so bizarre because I know someone that it's like a song getting started right now. They wanted to sound like a nice. Person.

And so it shaped it so much works. My clients. Always move forward like I did this. I never knew those words. No one really did anything about it. Let me it down to it takes two. They only have going on. Right now. Let's get stage to like when they feel like it's almost as let's say an organ. This here's the fourth day. Let's talk about race right now. Yeah.

Dominic: And your journey with Reiki has their proceedings I gave you a manual and it's got lots of different aspects to it. But I would be careful with number one how you do the achievements when you do it. Give yourself space to do the clearing between each and not to worry too much about diamond Reiki and past Reiki as many of these are ethical challenges and are not the original authentic Reiki as found in Japan. These are different features of the gold being which is the energy that you use in kind of like the gold ray and it's part of the kundalini thing you will get some symbols and things at some point. In today's session. I'd like to see if I can't activate your well of dreams.

Lily: OK What. Does that Do.

Dominic: You have dreams chakra is the one that's at the back of your head. It's the shape of a walnut and that it emerges where the base of the neck touches the head so. On the occipital lobe.

Lily: Oh my gosh this spot has been itching like crazy for the last two weeks right and it's time for them to kind of activate.

Dominic: It is a dormant chakra but it's the one that connects to the akashic records. So while we've in the beginning of our sessions together after the workshop. Last year in September we worked a lot with. First of all you didn't see anything. Or hear anything. And we work with feelings and we worked with love types you could feel the energy working and we gone through the blocks on your meridians and we got through the blocks in your chakras then we know do you try creams and things. And then somewhere along the process we got a lens wasn't able to get you there and you got the lens that could heart you see. Your Phoenix shakti and you could see the deity for awakening. [crosstalk]

Dominic: And you could see your. The books the tripod petals and lifetimes and things like that which is great in that vision is shaping up really really well. And more recently you're listening is up opened up over the last couple of months since a month or two or three as I've now agreed to do the basic toning of the different frequencies.

And we can be on certain levels and see in reality that you won't listen to the scrunching they do. And you heard briefly doesn't sound like you've been opened up more to that. So it really has a unique arrangement. I myself am different. I can listen to the delta a little bit I'm a little bit too my muse a little bit to and to the 26 dimension. By my listening isn't very great on the mental or The astral either very much. So there are certain levels that I could portray myself listening for if I wanted to but I've chosen to prepare to deal with every day where I give guidance that I need so you're pretty strong. That's when it comes to flowers and jewels. On the other hand you are for example is somebody who is very strong and have a mental movement.

Liliy: Yeah. Advice from the guides. That's kind of the pause.

Dominic: And then just tell me it's like talking to the ground to be able to talk to somebody on the physical three is under the ground you have to win this. Most of us in school that think there are special classes outside the library and you'll And with that in the way of dreams you'll actually be able to go to the library so the desk your staff the books outside the unicorn's the fairies all these things and have talks of them all go to the school that's out of the library. Do Classes like sacred geometry and I've shown you shots before like of the student sitting there levitating or energy duelling.

Dominic: I showed you different shots around the place but I don't think I've ever really anchored in your body it's about activating your Causal. Light Body. And the course of life body is really cool because you get to move around in the world where the Guardian Angels and the guides live. In the begging it can be a lot like landing up in Narnia or Water world. It is really a cool space.

Lily: So I've tried quite a few times to project to the library in really the shell and I can feel that some of my early years of using my body is not the same as when we do sessions and you help me. I never feel like I get more than 60 per cent of my energy up there. That would be great if I led to me being up there.

Dominic: That's exactly what it is.

Lilly: A thought just popped into my head about when would be a good time to integrate my higher self. When we really started doing that.

Dominic: When it comes to Higher self integration the best rule of thumb is whenever you're prepared to do the work. They already do in the heavenly realm. If you're able to do it they can join into your body and. Augment what you do.

Lilly: Okay.

Dominic: So is currently your higher self. We've spoken before. We mentioned that. She was a teacher. And she took She's essentially. At this point is she's like ta ave for the class. This does about healing about. Soul biology and spirit biology. And she's pretty happy.

She's there issues training. Briefly. And it could have been any time could have been a hundred years ago or a thousand years ago. The idea was to go through training as a serendipity. Which is the type of angel and I proved it to check my maths again to see where exactly does that fit in. It is like the half way it's close to half way up the hierarchy of angels. But higher.

And we mentioned before about five thousand five hundred forty-five different types of angels different ranks. In doing a hierarchy we could approach the book quickly and go through some of them. You know you yourself have done the exams in dreams or in waking moments in your in with your screen with your TV in your mind's eye you've done some classes to do with.

131

Healing with fractals and flowers.

And. They cover that syllabus and they're normally takes about three years for the person to kind of figure out how to use it and where to get their little certificate that authorizes them to use those skills that they're capable. And everything else. When it comes to Reiki I just be told that it's probably good to give you a message about. Being impeccable and being very careful.

Dominic: I'm treating it as a very sacred thing and not just give Reiki to somebody while you're eating food and your intention is distracted not to sit and do it while you're having a shower not to do it when you're swearing or something or just simple guidelines you probably know this already but this I'm saying it's important to use that in a manner. If you just get disconnected from it.

Dominic: To pray. To the energy you don't pray to the energy or to someone or a being you paying essentially to God or this space of like that or you know the intelligence is the creator or Saints behind things either you see it. And get reconnected to this particular be more energy Ray that has particular functions.

Dominic: At this at this point. I'm just checking in to see if all you all the achievements in place that are in place you can't get of. People often ask sometimes they're going to go to the energies of flowing any-more and I do it can get very attuned. And the answer is no you can't get triggered once you made that connection you've made that connection to reality with a beam's the rays originated and the correct way to get connected again is to pray and be sincere about. Have you learned your lesson to you who has the energy to try to do something wrong somehow figure out what you did and then go back.

In this case I'm just saying your relationship with the green ray seems to be using it everywhere. It's a possible fault for everything. It's actually meant for mental healing only healing of mental events in the person's mind on the mental primarily the white ray just to remind you. As for healing on the emotions of bad it is. It is much more efficient when they can do emotional stacks which is what I'm going to teach you in today's lesson so emotional stacks give it something to act out. It's like stacking up emotions you are holding like a block where a pyramid with these feelings and then is able to work through them. If you set up things correctly. These things can work and they can be amplified hugely if you you're right halo's those that match the colours that relay one to heaven. [crosstalk]

Dominic: Then also when you read what a lot of functions work with the Green Ray there is a system. So if you grab the kind of clues the puzzle like this and then is the purple halo which stores symbols, chants. And geometries. So you don't have to cast them all the time. Does that. It also works in space and time like this and then the crazy thing about emotional bonds. There's a number of things.

Lilly: So it took me today to meditate. Did you come to check in on me?

Dominic: Yes and when I spoke last week and talked about the different journeys that symbol would appear in front of you when you were ready to kind of become Reiki. The idea is that you be so in tune with it you understand. Some people have sons have dreams being been through a maze. Sometimes it's about going to understand one is energy that really do.

Lilly: Yah I was busy mediating and I have the image the image of your golden light body. It was you definitely you were smiling and waving. Really happy about something.

Dominic: Often I step in when students are doing healing, microsurgery or they need some help with me to do psychic surgery with somebody and they're just busy training and they've done some practice with me or something. I often go to go out on the Astral and I can copy myself pretty well at this point to work with multiple people and to act if I'm allowed in their premises and obviously we sign a contract to say you know he's allowed to come in when we do hearings and this and this and this and this.

And if your client's contract includes that just say. You know. That I can assist them or other entities of my own. How are you going to word a bit if you can put it in there somewhere that you can come in as well to kind of. Help. With the healing session if it's necessary. And so I've got a couple of clients overseas who have stress relief practices which is the same as having a medical practice. I can actually do psychic surgery at their practice with people like. Or doing some doing helping them with getting rid of tumours and things. With paramedic pylons. And redoubling getting the tissue and doing lots of advanced things with healing. Many things I can do with them often aren't doing that. Sometimes it's night-time and I'm aware of that at night and I say OK I'll do this for now.

Dominic: Only for the guys who are not in the States you know the time will be. An early morning sometimes if it's for people in India and so on and so forth. Yeah I've got a busy schedule with students all over. So basically Sometimes I'm there and sometimes if you tap into the gold what you were doing is you were tapping in to. The gold the Buddhic level of intelligence which is what you want to with your golden halo. And you saw briefly the temple and then some other Buddha's and then me and I kind of smiled and winked and I knew those the students of mine. That you can hear what they're doing what are they doing what's visually what's happening on the planet what can I do. Like is that you've got an interest to have with children or just to work with animals. And I think you've gone with. Old people and you said you wanted to go with the kids.

Lilly: Mainly kids and animals at this point.

Dominic: OK. So you might want extra at some point to work on the body golden listening. To figure out what to collect. Sometimes everything like the and the Sanskrit poetry and you know different lectures and mostly these days is they're talking about the love types the use of the Love-type Haversheim frequencies for healing and for spiritual about and in awakening. [crosstalk]

Lilly: I feel that is going to be every two months.

Dominic: See other condoling new Reiki it's going to take in the different places new dimensions. Realities and plane-walking. You would think that once is as it will do this as well but it halves with. You travelling and this is a new kind of trouble and the heart's with plain walking going.

Dominic: So you could be travelling and these and these and I remember when I did the Reiki and in Kundalini two or three it was around the same time I did this mass of Crystal grid going around myself and I was silent for four hours of meditating through stuff and it was like I went to a shop and a book bracelets and rings and I travel to the library and I travelled here and there and I went all over the show. It was only later on that I figured out what I was doing with this thing. It really is about showing you about the different levels of creation and levels of being. So it we'll pick you up and pick you out of your out of your reality and do that kind of thing to you.

Lilly: Strange that you would say this say that those morning throw I'm thinking of making a crystal post like manifestation. Good morning I finally got stuff.

Dominic: That has protections placed in it and also us there grading Chris like this 24 hour days shields and things are either taught you for like energy from being slung from you and one for bucking mind reading and one for blocking pain transferred to you in sessions that you have with other with your clients.

Dominic: Basically you know they accept crystals don't accept anything I was say fancy words salad that's really nice. But you need to is. There's a fire in geometric terms. What they must do. So. I've been I've been planning this Christmas good for my own house and I'm thinking what I need here. Is to protect from. You know different things to many spirits coming alive. And there's an air happening or something at night or whatever you need. And then a couple of different examples like protecting the cold from being cast and protecting. There are a number of different things that we could look at. Briefly I will. Mention.

Dominic: That you probably know already but it's useful to kind of just recap. OK. So basically that is one for the mind reading sheet the brain reading shield one foot to for blocking the emotional transference of pain one foot locking theory implants one for blocking attacks on the theory or the physical level the lower physical one foot being struck by the energy. From being stolen to ward of negativity to ward off attacks from other people with. And that's something that you will have had would have talked about them the ones that you needed. During our time together so you have them written down in your journal on a book or something like that.

Dominic: So. Why do you want to deal with. In today's session. Do you have goals and that we are working towards. Other than listening and maybe we are activating the course. Maybe let's start with that. I'm going to give you that I really give you the actual sacred geometry that you use to do this with somebody so that you can do this view of self. Sometimes with some people depending on this contract I have with them.

Dominic: Some people say they don't want me to touch it with any energy at all they just want to be led through it and do all the work on themselves. Although I do it all the angels do it or someone does it. Some people say you know it just depends on peoples boundaries and so on. I respect all of that. I'm going to show the geometry and if you don't mind I'm going to dislike the thing your dreams actually has been kind of like you said it's been that it's one thing to activate. That's why I picked up on that and I think it's okay. Let's do that. To be opening up for a reason at this time and you're well prepared to kind of handle. I've seen some shots from the from the lower courts of the lower line of the six dimensionally. But you have read what goes on there. I could move around there and do some stuff. So let's end you've been very patient also with me.

Dominic: And then you draw a white circle inside inside that circle. You can visualize. Now the well of dreams is open. You're aware of dreams which you focus on the back of that just in between the occipital and your neck.

Dominic: And you might feel like there's like a kind of like some people have bumps in between their eyebrows where their third eye is down. You might even have a bump in between your eyebrows as well where your third eye the gland for that is the particular gland for that one is.

Lilly: This has only been aching since I've had he Reiki attainments.

Dominic: Kundalini does things and it gets you slowly towards progress to going by getting through your clout and getting through your baggage and the proposals one through this incredibly Reiki is even more like that. And it's just like riding a riding a wave.

Lilly: Yeah it is. I dot even have to try. People have said to me, you've going to burn out.

Dominic: It's. Pretty big energy boost. I remember it and I was able to deal with so much I could deal with a bad time I was out of the house and I was dealing with. Running a sushi bar. And I had human clients and I had like a pope. Obviously the red and my design covers for books and I was doing a number of things. And I was able to handle. It spiritually there was a lot going on a lot of awakening. I kind of remember that time where I was there and I was kind of like yeah I'm so close to kind of going or jumping off the cliff and fly.

Dominic: And it was there kind of time.

Lilly: I realize yeah.

Dominic: That was I was able to deal with so much more business ever so much more energy with the current kundalini adventure with the alleluia catching the phases. You get kind of like a five-year plan a three-year peak and a six month peak. And then it goes in and after five years it goes in two.

Dominic: I wish I could sell you a tremendous stage where you can read in the book called biology. of kundalini by Joanna Dixon. Which is recommended reading. You can read about how it affects all of us that hard grows the kidney's hydrolyse releases adrenochrome which makes you feel paranoid possibly and what you can do to deal with the hormonal and chemical effects that are happening in the body the cascades that are triggered by the continued enlarging the heart and doing all these things over a number of years because remember it takes four years for heart cells to grow. It takes a couple of years. There's like a seven-year cycle and then burn out recovery. And Death of the ego depending on which part of you are on Shiva pat or Shakti pat. You are on Shakti pat route. [crosstalk]

Lily: What is that. Authors name again.

Dominic: That's Jana Dixon. She's a lot like you.

Dominic: Do you type in Biology of Kundalini into Google you'll get the website. Read her story. It's a very exciting story by meeting Mr. universal. She has this idea of this perfect man. And she imagines him and she has you know orgasmic times with him as she does all sorts of things and then one day won't spoil the story but she arrives in the states somewhere. And she's going to a book launch on spiritual matters and she finds him and he's real. It's there's an exciting book it's written for a woman. I read it anyway. But lots of details fall of medically inclined people who these people really exciting.

Dominic: Would be. And the thought process is very proud it's very beautifully written and it was gone scientific and everything.

Dominic: OK. So let's see. How are you doing activating your well of dreams?

Lily: You know is this something I'm suppose to be doing. I can feel it vibrating.

Dominic: Yeah. Vibrations are going out. It might take a couple of minutes but a coupler you'll notice. You will notice that you'll be sitting at a desk with a screen and books and books around you and floor and in their desks next to you. Then you see me the distance wailing at you. When I find where you park your location in the library.

Lily: Looks like a lot of books.

Dominic: I remember when I was a little a little kid I sometimes had these dreams and I were going to the reading rooms in the library and I would explore and all these things would happen and I'd meet you know different people. And that was really exciting. Remember from the causal or your parallel selves. There's a room for music full growing you're listening there's a room for seeing the futures on this terminal. By the way if you'll allow me to kind of pitch up here you use your library card which is blue. You swipe it across the top of the screen. And when you swipe when you swipe it across the top of the screen you see it opens. The pink card is for viewing scenes and five times around you.

Dominic: Yellow card.

Is for a timeline room is where the parallel one is for seeing the future and you'll see that on the screen you've got like. A triangle with tunnels in as you can see a future handless see where you're going. You can see there's this vision of yours is much clearer than the Upper Mental is right.

Lily: Yeah.

Dominic: So you can see it as you go. You'll feel vibration still in the centre for a little while but. Basically. As you go along that vision will get better and better and better. So you can actually start to see time tunnels and all these different things.

Dominic: Dream is from last week or what have you been doing your angel dreams and Caribbean training all of those sorts of things. There's nothing to tell you in the library. The planetarium with all the planets in the solar system is enough to entertain you in the library for many many many many many many many months. At some point we'll venture out and go to the school and attend classes. Are you doing them in couple of months from now so weeks or days you kind of at the end.

Dominic: OK so it's like that.

Dominic: Is just simple. Shay does that it's amazing. Yeah. Okay. So basically. Now I'm going to show you how to do emotional stacks. The main teaching of today's session.

A certain friend of yours needs healing. And you don't often reveal names and things to people in sessions and all that. Oh I call into different names sometimes you know. For various reasons I'll just protect my clients privacy. Names and changes and all the rest of it but I sound like someone says that you can demonstrate with their chakra or whatever it is. And then we go ahead. But she said no. She says that you know it's me. Thank.

Dominic: Can you see in the back heart chakra. It's got some of it is in enlightened and a little bit. Then it got some pedals that are not. Doing too good.

Lily: Especially in the two and three positions.

Dominic: So last time we compared the use of Reiki and simple force and light types. hat do you We used petal force and 5 chants to heal the mosaics. And you can use the big mechanism I taught you to heal generally. You kind of want the yantras going to be blown up anyway by the continued caution for the center of the chakra. The varga or the yoga mats in the throat chakra and the heart as these little hearts similar sort of things and they need to be repaired with courage force diagrams. Then you've got.

Dominic: This time We're going to fix the chakra and then we're going to use the same white ray energy to work with stacks. Of emotions until you get an idea of what it's going to be like. Okay. Try for me Reiki one if the yellow energy. Think about the symbol for Reiki 1. And you can draw an out in your mind. I'd put it on the side of your screen. So to the to the left. We've got your spiritual reading list to the right of your table. The first book is called: You know heaven. The second with a yellow cover is where you must ready from for next week.

Dominic: But you can see it's turned it over it's done it's done interesting and it's done a couple of work. Reiki one by the way turns the petals over and golden them. While petal force diagram produced a mother of pearl colouration.

Dominic: The physical body the thick vortexes. The Yellow ray of creation doesn't do anything for. The mental level or the or the course or anything else. It does communication for you and your higher self. It does communication for between you and maybe other people. It connects two things. Consider as do anything else and it does a lot of things we I'm saying just in general for those dimensions to be concerned with for have it does a number of things. It does like. I think I made a list once of like 17 things he does for humans and the rest is still up. It's a mystery. Probably there's a lot of things for healing different kinds of forms and different levels. Sure no problem. Before I make a mistake of saying it doesn't do any.

Your external for external vision for seeing more eyes is getting much better. I did explain to you about butterfly eyes take about seven years to see things it first works with Yellow where you walk.

Dominic: And then you see this. Yeah.

Dominic: Yellow start starts with yellow and then you get.

Dominic: Purple.

Dominic: Yeah four hand movements and things like that and then you get the blue for indication which means like somebody is in clusters indicating this person is talking indicated and you see blue for colour and things and stuff on that level on the etheric level certain colours it easier to see and meet different things. Then eventually you'll get to see frames and you'll get to see the stuff had you feeling like you feeling like now in your head you pushed up a middle.

You can actually feel a frame. With stuff on it. That's what we were looking at the oval frame the mirror to see pictures on. Now using your library lens to see things. And the wonderful thing is that you know you can actually feel mental bubbles and people's chatterers opening up and before the pedals. You do all of that now. You'll get to maybe start to see some of the stuff that you're feeling. Which is really really cool. For better television when you fully develop your green eyes or your blue eyes or mermaid vision or whatever it is it's some kind of vision on the ethereal. It'll give you some. You can bring it down the bubbles the bubbles of people's thought bubbles to that s to the theory. You can bring down emotions to the ethereal. You can bring down frames from the elemental to the aether to see what's going on in black on Skype to someone or the gurus do this sort of stuff and it's really kind of cool to do that. So I think last time we gave you.

Dominic: I have to use some symbols with fire seven for activating a Kundalini in somebody's case. One to do workshops in the future. You probably will because it looks like women in your future are reading and you had a look yourself and you fell through it. You saw out work doing the workshops really helps.

Lilly: Yeah. OK.

Dominic: So do we actually it's exciting time where we've activated this thing and we now are working with. You've done that macro. Now. I'm showing you some colours.

Dominic: Now these on the physical body because you must know that once you've done that and you're track list you must keep up your heart and you must keep up recycling stuff and not suppress feelings and all this kind of stuff. So now I'm going to give you somebody else's body who has volunteered for this clearing and they're ready for it and they know that they have to do tapping. But they're busy with this process of clearing nadis and so we're going to have some time with them to work on.

Dominic: Pulling out emotional stacks you can see on the screen you've got these lines which are these parties over this figure and you've got. Yellow stress red stress six months black and grey grey stress maybe a few years. Back probably from a long long time. Eventually ultimately you should see this thing glowing with vital energy moving up they've had that condoling the awakening they're working on this journey with me to help them get through all of this and I'm asked in advance. If could use them.

Dominic: So here we go that they've done like 70 percent of the work themselves so I thought yeah this right they've done a lot of them by themselves. They can get a little bit of heart. And there's no problem. This is where we can kind of pay it forward so we read to me a while ago you said how can I have paid for it all. But isn't it sounds record they're doing. Why do we do about other worlds issues.

You'll get into that when we ask where you start listening in and being a little bit late at a little stage when you feel ready and will probably be a couple of weeks from now two weeks to a couple of months and that's fine. So let's see this whole idea. Just watch the library. You watch you next to the monitor now and your vision for what's happening next to the monitor is like 20 times better 10 times better vision of what you have inside the monitor is inside the monitor is the upper mental.

Dominic: The middle so obvious it's the most causal but it's mainly that. And if you look outside the screen not to your desk outside next to there's a device that plays C.D so you can bring in a seat from other people's dreams and you can slot it in and you can play the dreams on the screen to yourself and see what's been going on with people. This gives you a lot more tools. This shows you what are you working with police you'll be like are you hearing all these things and I'm like you know when I'm going to say. OK.

Dominic: Well today is your lucky day.

Dominic: It's happening. OK.

Dominic: So he began if you can rush to the right of this group I'm going to show you how to reconstruct. Emotional stacks and you can watch the screen occasionally to see what am I doing. The person on the other end it is now. 10:00 a.m. for them so they're busy tapping.

Dominic: OK. So here is the emotional stack. I'm sorry inaudible like a. Block. And next to each of the glass on the outside of the block you will see certain symbols. You will see that of God like the Will of God little triangles. You'll see that they're basically the shepherd that holds you last week before the week before about hope love related I've never recovered them again today. So you know them again to clear emotions on a certain level. So you build. These little boxes. You see as the bottom. Right top left side. And the front and inside each of these boxes you might see that you've got these little force diagrams indented into them. Energy Is jumping into the box. Is jumping into this box. And. You see it right create s nodes or dream building.

Dominic: Another one you might hear with your ears like is a ringing sound which is so it is salvation love giving you listening. You might you like a. Like that white noise you might feel like a ding ding ding ding ding ding ding or something like that. Sense at every corner of that box that you build you'll get like a sense of

145

oh you must be doing something on a spiritual level or something. If you teleport something I'm very busy in my well the pulling students souls around the show.

Dominic: But you can run. Oh my goodness it is actually happening. It's it just my imagination. Well I used to think it was chakras balancing up or something. And I know that's awesome. That's a sign to know that I'm actually doing. Something. Now it is different. Different for different kinds of love types are two different things have been sequences of different kinds of sound with the best kind of thing is kind of like a shimmering type of sound. OK. Would you like to construct one. Can you see the boxes to the right hand side of the monitor.

Lilly: I can see in a circle vaguely. The bosses are doing separate voices. It just looks like I should see it was outside of that you are going through we will see you in the black. Boxes that you mean. There's a saying no to.

Dominic: We've constructed these stacks. But they're basically what their function is is to hold and to contain and you must watch. It's like if you see the difference and we probably should show a difference before you can do it before and after anyway. If you use white ray of creation and you might want to draw the symbol out. Can you see how it just moves through that box?

Lilly: Mostly light sort of from the inside.

Do you see that on the screen you can see that you actually move? A lot of grungy stuff out of that profile that's on the screen.

Lilly: Yes there is a lot less yellow and less grey

Dominic: Try using Reiki on its own without a stack and see how much you can do.

146

Lilly: It is moving stuff our but The stack is faster. Yes.

Dominic: That's what I wanted to demonstrate today. I wanted to see you really. I think it's a bit silly to put it in a box and then the Reiki through it. But I wanted to show you which one is reading a bit of an experiment. Which one is. Going to work stronger. You might thought Reiki by itself is enough

Maybe you know maybe it is. But I found it interesting. It's just you can't prepare things for your energies and words are super charges that.

You a of this is a technique to kind of supercharged if you just use them like gently. Go through somebody's energy and emotions just a little bit. Use there's just the white ray. if you want to more aggressively have them recycle stuff get them to do the sacred geometry the herb love the clarity love the serenity love and everything else that's in a kind of work with you with that courage love and so on. To prepare the chapters and do all of those things get them to use it as the next phase. And then if you want to if they're really ready for it you can use these stacks to supercharge that.

Dominic: Kind of set up. Better that can work much more efficiently. OK.

Lilly: So there we go work smarter not harder.

Dominic: Would you like to try to make a stack and then have a look and see how it's how you do. And then you can have a little bit more from this individual maybe focusing on the yellow to get a yellow and the red. So for that. And as you as you need you're just trying to make a box. You put your shoulder really pretty school. It's a skill that you're learning.

Dominic: Symbols might look a little bit different to the one that I use in my box but it's when it is working for you as you familiar with to hold the energy and to recycle. It's usually types on the outside. So this is why the basic concepts basic and sacred geometry is a useful thing to have. Over the next couple of weeks. I will be introducing you to a sacred geometry class where someone asks can then take you put classes whenever you go up to meditate during the week and they can assist you with the basics of sacred geometry for healing. We've covering anyway a lot of the syllabus that's in my classroom full for healing anyway. But somehow. The basis is a separate jumble is going to harp you with coming up with your own symbols for doing things for grades. It's the language of God creator in this omniverse. Can you see that you've made you've made. Three sides made a bass. Then you've made two sides to that. So now are we going. You've got the you've got it and you just hold it. Sometimes you need two or three to hold everything but sometimes I'm just like.

Dominic: I just make one an extra one and it inflates the big one that gets bigger to handle more energy. But most people most people will probably take a while. But you know it really starts to have them in it. The thing is it stays. It stays active until that energy has been filtered through. It's almost like setting up a grid of crystals to remember in two minutes they're putting a protection on something every morning. I was clever one day. I'm going to put in my necklace and leave it there. You get a bit lazy but do you know why it's the same thing this stacks it's there. And it fills us up the energy and the Rays are very intelligent.

Dominic: So if you're working with multiple people I work with multiple clients that box can each have a little box at my desk for each of them on their on their desk. So it's their fault is their process for the next three days after the session to get them through trip through that energy because it takes about three days to get through all that stuff. Anyway you've lived through that experience.

Lilly: So you know this is behind the scenes but I was doing to help you. And I agree that really nice you see.

Dominic: No problem. But you've gotten to the library and you can go to school quite easily if you want. I can even drag you the sacred geometry class. Now already or you can start your day.

Dominic: You might find that it's difficult to move. I should have told you that you could use and conditional love which isn't which is a pink semicircle. For you that unconditional love is have things move on because anyway it works like that for causal bodies. If you were stuffing you couldn't move your head cause the hand and stuff you would have had to sit down.

I can't move and you probably won't talk to me though. Okay I get that I can see everything I can't move. That's how you get out of it but you look like you could move. So I thought okay she should probably be fine. And so you treading the backs of your hand and you can sense that your energy is in that body rather than in your physical which is where the energy was going the whole time. So now you know are you seeing in you're able to be aware of what's going on.

Lilly: That's. So cool.

Dominic: True. Next week I'll take it as sacred geometry. This is maybe another trip for today. You can come back if you want to your desk. It's going to be easy. You can practice growing up and down projecting to the course should be much easier. And now you're right you're well have dreams active. And then you start remembering your dreams more and all sorts of things to start out. That's a huge it's a huge thing it starts to hit you. But you were right before it's so no problem at all. That's it. OK. Do you want to make some time for next week?

Lilly: Yes please

Consciousness

EGO

Personal Unconscious (2D)

Collective Unconscious
Dimension 1

3rd dimension – Ego

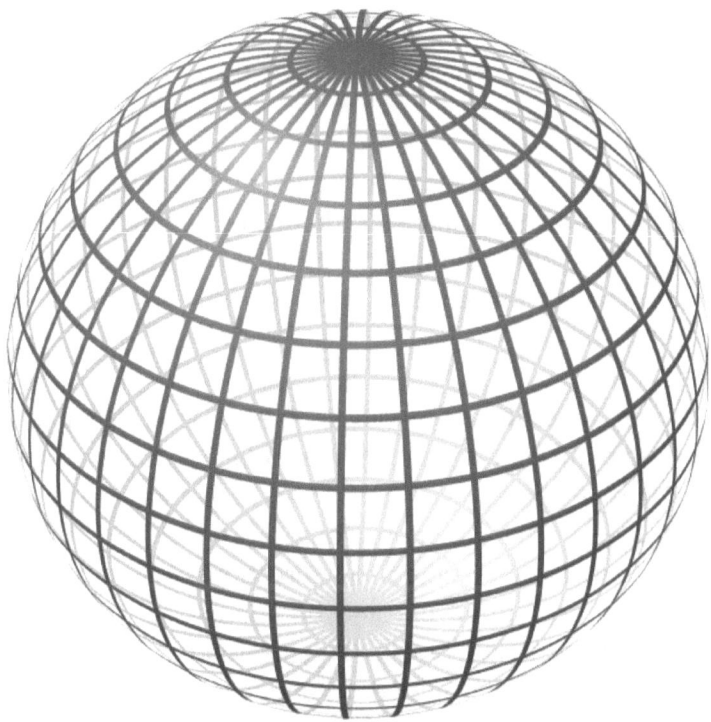

Illustration 11: The easiest psyche element to spin - the ego (typically cleansed in a sky dive manner)

[I want my powers]

All bad qualities centre round the ego. When the ego is gone, Realisation results by itself. There are neither good nor bad qualities in the Self. The Self is free from all qualities. Qualities pertain to the mind only.

Ramangwana Maharshi

[Use Blue, Red, Yellow and White Reiki energy to clear.]

Our universe has three spatial dimensions: length, width, and height. In formulating the general theory of relativity, Einstein showed that time is another dimension. According to the general theory of relativity, space and time communicate the gravitational force through their curvature. The special theory of relativity is Einstein's law of space and time in the absence of gravity.

In 1919, the mathematician Theodor Kaluza unified Maxwell's electromagnetism and Einstein's theory of general relativity by adding a fifth dimension. Thus Kaluza was the one who suggested that the universe might have more than three spatial dimensions.

For example, a garden hose viewed from a long distance looks like a one-dimensional object. When looked at closely, a second dimension, one shaped like a circle and curled around the hose, becomes visible. The direction along the hose's length is long, extended, and easily visible. The direction circling around its thickness is short, curled up, and harder to see. Hence spatial dimensions are of two types: large, extended, and therefore directly evident, or small, curled up, and far harder to detect.

As for the garden hose, the curled-up dimension encircling its thickness is detected either moving closer to the hose or using a pair of binoculars from a distance. If the hose is as thin as a hair or a capillary, its curled-up dimension is more difficult to detect. In 1926, the mathematician Os kar Klein applied Kaluza's theory to quantum theory, which is used in modern string theory. Klein showed that our universe's spatial fabric may have both extended (the three spatial dimensions of daily experience) and curled-up dimensions.

The universe's additional dimensions are tightly curled up into a tiny space, a space so tiny that it has so far eluded detection.

These extra dimensions are believed to be minuscule, somewhere between 10-35 meters and 0.3 millimeters in size.

The equations of string theory show that the universe has nine space dimensions and one time dimension. At present, no one knows why the three space and one time dimensions are large and extended, while all of the others are tiny and curled up.

Symmetry is a physical system property that does not change when the system is transformed. For example, a sphere is rotationally symmetrical, since its appearance does not change if it is rotated. In 1971, super symmetry was invented in two contexts at once: in ordinary particle field theory and as a consequence of introducing fermions into string theory. It holds the promise of resolving many problems in particle theory, but requires equal numbers of fermions and bosons. Thus, it cannot be an exact symmetry of Nature.

Supersymmetry, a mathematical transformation, is a symmetry principle that relates a particle's properties of a whole number amount (integer) of spin (bosons) to those with half a whole (half-integer or odd) number amount of spin (fermions). Bosons tend to be the mediators of fundamental forces, while fermions make up the matter experiencing these forces. Bosons can occupy the same space and have an integral spin (0,1, .), while fermions cannot occupy the same space and have a half-integral spin (1/2, 3/2,.). For me super stings theory is the begging to define and make tangible the substrate that is consciousness.

2nd dimension – Chemistry

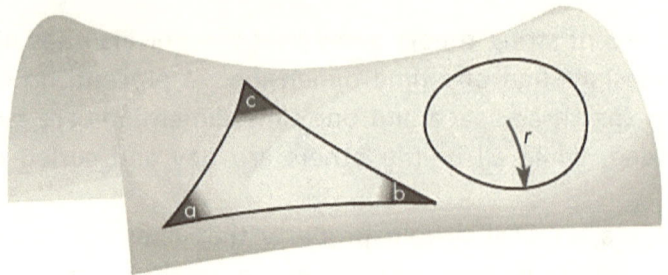

Triangle: a + b + c < 180°
Circle: Circumference (C) > 2πr

Illustration 12: The rotating flippable primordial atoms

[I use Red and yellow Reiki rays to cleanse this psyche aspect.]

Bosons (compressed souls tightly wired selves are consciousness) transmit such forces as photons, gravitons, W and Z particles, mesons, and gluons. Many bosons can occupy the same state at the same time. Fermions (e.g., electrons, muons, tau, protons, neutrons, quarks, and neutrinos) cannot share a given state at a given time with other fermions. The fact that fermions make up matter explains why we cannot walk through walls: the inability of fermions (matter) to share the same space the way bosons (particles of force or energy) can.

Supersymmetry treats all particles of the same mass as different varieties of the same superparticle. This means that there is an equal matching between bosons and fermions. A supersymmetric string theory is called a superstring theory. Introducing super symmetry to BST engendered a new theory that describes both the forces and the matter making up the universe: the theory of superstrings. A string theories are different aspects of a string theory that has not 10 but 11 or more spatial dimensions. This was called M-theory. The M might stand for the mother of all theories or mystery, magic, matrix, marvel or membrane.

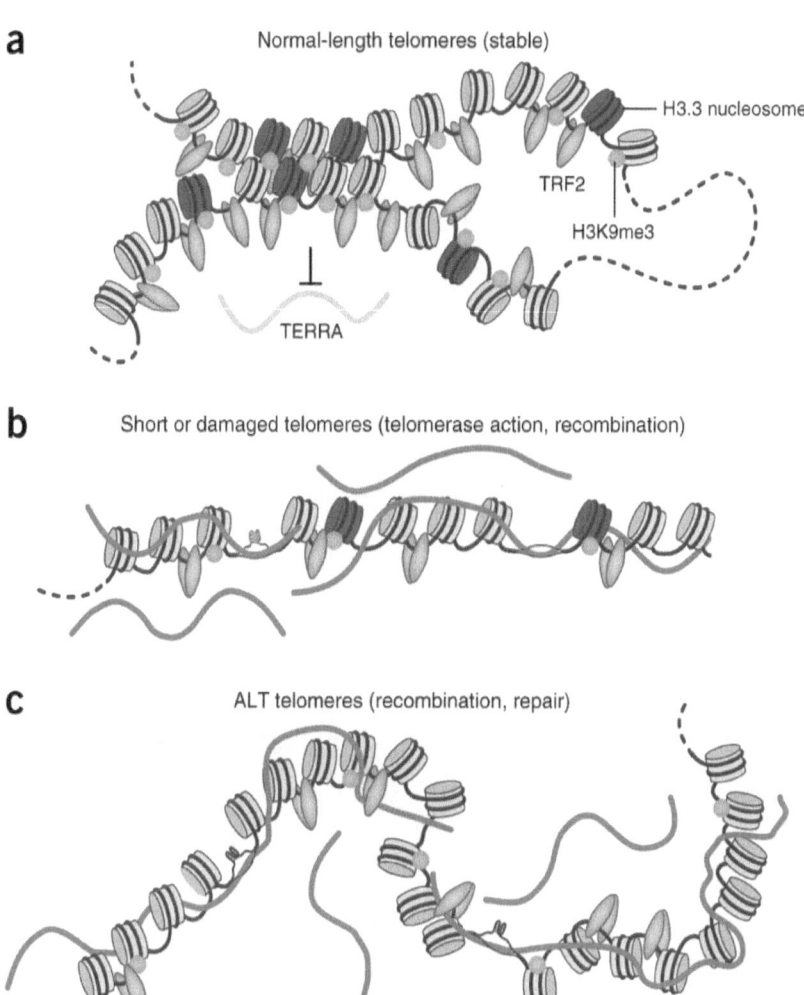

Illustration 13: long telomere's mean good health

From DNA till Sensei level

1ˢᵗ dimension lambda (DNA)

We use Gene lineage previews, double helix water therapy and cancer screenings to heal short telomeres (a to c) in our very sacred DNA: As echoes in the works of renowned spiritual leader Credo Mutwa and through listening to talks about Indigenous knowledge, Illiterate Mr Johannes Mati taught me everything I know about Ukufa KwaBantu and African Cosmology. [crosstalk] Use Yellow + Grey Reiki rays.

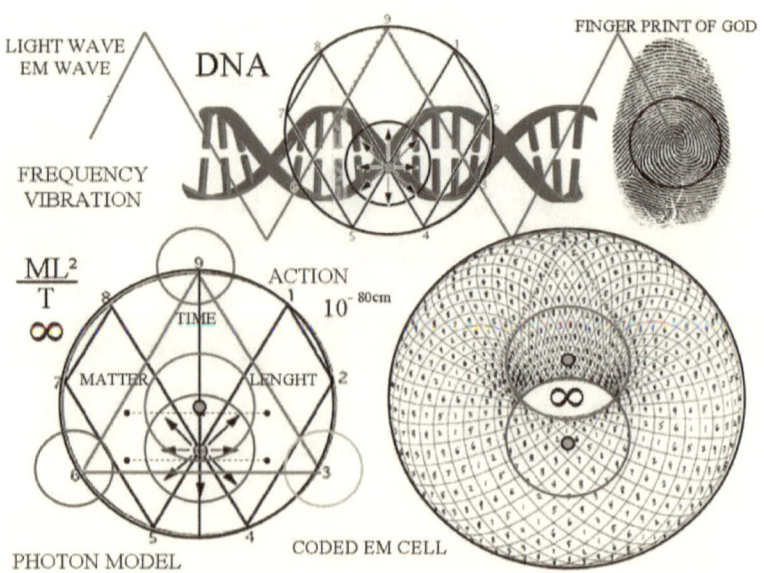

Illustration 14: Comparing the particle and wave model down to lambda

In late 2013 I was contacted by, Mary Elizabeth an out-patient in partial remission of liver cancer. On meeting we found this most marvellous resonance. I had enjoyed success with cancer before particularly in complementary clients and but this case would prove a pious one for any scientist left in our group. Mary's son, a medical technologist, would be watching the cancer on this MRI screen each week.

2013/04/17 05:33 PM

The chemo is stopped this week. It is two weeks on one week off. However I was off earlier, because my feet were so bad, skin wise, I could hardly walk. That was my resistance to walking this path, still... I am working on that. The coming Monday I will see a new doctor and then probably go on it again. Any help is appreciated as I feel I need/want the spiritual component.

2013/05/08 06:30 PM

You came in through the solar plexus and seemed to take inventory of the spiritual state of the body. You probed the solar plexus, heart throat, crown and then went back down to the rest of the body. You checked on the liver (cancer). I felt you go to the solar plexus and you pulled some fear out. It was resisting, but left. It was transformed into bright, positive energy and corkscrewed back into the solar plexus. It went very deep.

I felt you check out the aura and it felt great. From around the center/ slightly under the breast the rest of the activity took place. There was an explosion of electrical current from the heart(?) about ten centimetres under the left armpit in the main

body, huge infusion/opening of energy. Vibrated through the whole upper left side of the body.
- Second one vibrated from the same spot through the whole left side of the body.
- Next one vibrated to the belly, scar area and solar plexus
- Then the whole right side of the body
- Two more smaller ones followed. I went into bliss: body and aura.

Feel great rejuvenated. The opening is still activated.

I loved the feelings. Thank you again. And thank you for the message about the cancer. My feeling is also that it will be resolved soon. The medical info, blood work, is all very positive. Scan on Friday.

2014/02/21 11:47 PM

Here is my update:
- still see newness in snowflakes, lights
- Heard very faintly the Nat. anthem of the Netherlands in my right ear. The words just flowed in my mind. Interesting! Orange on top.....
- realized that the chaos from the cancer was an opportunity to release fear by me and the family
- I hand out happy cards and the message needed to be changed. The urge was very strong, something new!
- surprise, I could feel my spiritual right leg moving separate from my physical leg. What a feeling!
- Was at a show yesterday which was very loud. Life music. A small piano piece stood out for me and touched me.

My cold is still there and I was wondering if we need to do more to solve this??? It makes me tired.

2014/07/12 08:28 PM

But it is a good reason to give you an update.
I am doing very well. Cancer is still gone and will remain so. My blood pressure is going down. I had an intuitive message that it would solve itself and it is doing so. About a week ago or so I felled something huge leaving and I feel different. I am gluten intolerant and feel so much better now!

All feels the same yet different. You really surprised me by saying I was partially and fully enlightened. I did not see that one coming! Life has moved on mostly the same, yet there are many subtle differences. I was surprised at your sudden departure, yet I realized that my greater self probably told you to do so. About a month ago or longer I was meditating or in bed, same room, and there you were, a huge presence in warrior mode. I knew it was you and all felt resolved.

I am having a closer connection to nature. I can feel its presence and gentle exchange of energies. Really beautiful and more intense. My self is telling me to look deeper insight and find more connections, so that is what I am doing! Intuition works a lot stronger, which is really helpful.

2014/08/08 09:54 PM

I hope that you are having a delightful day. I am. Today I am on cloud 9 as they say and loving it. I followed your advice and communicated with a flower that was particular beautiful. It threw its energy in me, as if to say finally you are listening (my interpretation of what happened).

Since then I can feel the energy of nature very strongly, not individualized, yet, but as a whole and I am thrilled with the connection. More and more animals and insects are showing

159

up in my garden, which is less beautiful, because rabbits and others have eaten most of the new plants! I see amazingly big butterflies in beautiful colors, hummingbirds and colourful birds.

I am driving on the road a few days ago and I see two cranes, lovely! I thank the universe and then the canopy of trees over the road changes. It looked the same, yet it was all energy! I hoped I would stay on the road and just enjoyed. I was in a totally different space and I felt the connection to nature deepening.

On an overall note I see newness, that what was old, looks so different. And if I do not see it I just have to concentrate. Life is full of newness and wonder. I am having a blast.

Log light Colorado: wax on wax off

I am at my end looking for ways to care for Sybil during her relapse after being n remission from the radiation. After a little bit of research I come across the loud telepathic sending of Maggie Appleton McBrein who introduced me to Ben Okri. A cancer surgeon at the most Saintly hospital in Colorado. To continue my training in log light over what I have achieved in mechanisms. That he would take care of my training.

St. Josephus Hospital in Denver Colorado. I did the short online course for para medical professionals and flourished. Each time practising on furling flower after flower of the Fibonacci spirals that I was assisting Syibbl to heal. That a sequence was formed by the patterns made by my voice. A tough thing to do she had already taken the death trip of radiation therapy. It clears to down to the bottom of the bubble of this omniverse. The wrong direction: down.

Maggie a plastic surgeon - without the plastics - an icon in her time, planned to take Sybil to China when she went of a rotation, and should they be success full in acquiring the support of the monastics there. Find full remission. Through music therapy and chanting. Sybil made the initial 6 months and, and though plenty happened, her life force the head monks explained was out. Like Ben would later remind me: we are matter waves made of bulky cadenced lights. [crosstalk]

Summary and Conclusions

String theory gives a theoretical description of elementary particles and treats them as one-dimensional curves (strings). Traditional models of interactions between elementary particles are based on quantum field theory, which treats particles as dimensionless points. Theoretical physicists have not developed a workable theory of gravitation that is consistent with quantum mechanics' principles. However, treating elementary particles as strings permits the derivation of a quantum theory that encompasses all four forces.

Super-string theory, a combination of string theory and super-symmetry, treats particles as very short (10-33 cm along its single dimension, which is 1020 smaller than a proton's diameter) closed strings (string loops). All of the masses, charges, and other properties of elementary particles result from the vibration of these superstrings at different frequencies. The complex mathematical basis of superstrings involves 10 dimensions: 9 spatial dimensions, 6 of which are invisible, and time.

Since superstring theory provides a unified description of all elementary particles and fundamental forces, it is sometimes called the theory of everything. Some major unsolved problems of string theory are how to condense, 10 dimensions to 6 (spatial) plus 4 (space and time) dimensions, and what is happening at distances smaller than 10-33 cm. In addition, the experimental verification of the existence of strings in the near future poses quite a challenge. Since they are thought to be less than a billionth of a billionth the size of an atom, we cannot use current technology to detect them directly. An indirect test, however, will be carried out within the next decade or so by the Large Hadron Collider, a huge atom smasher being built by CERN (European Organization for Nuclear Research, located in Geneva,

Switzerland). There also is an urgent need to develop new mathematics in areas of Riemann surfaces, algebraic geometry, singular geometries, number theory, and other related fields.

Could the Arrow of Time reveal a Strange Attractor?

The butterfly-like, two-domain shape of the Lorenz oscillator is well-known in deterministic chaos. It indicates a specific kind of strange attractor. This paper and slide presentation suggest the possibility that our universe is itself a double bubble oscillator in a dynamical system, and thus a specific kind of strange attractor. In some ways, our universe is like a giant Lorenz system, for it oscillates between two polarized domains on a path iterated by what amounts to three coupled, ordinary differential equations (ODEs). On the other hand, in some ways it is more like a Mandelbrot set, for it contains a whole set of ever-iterating fractals that evolve their contents within the basic form. Call our universe a DB attractor.

Regard the two domains of this DB attractor. They are the huge twins of our Double Bubble universe. The upper bubble holds space-time with the properties of 3D space and arrow of time, plus matter, energy, and our known gravity pole. But the lower bubble is a flip-flop mirror domain. Its time space has the counterbalancing properties of 3D time and arrow of space, plus antimatter, tachyonic energy, and the lost pole of gravitation.

(It's important to note for later: this TOE proposes that each bubble's arrow of dimensionality, either space or time, is really only ½D, since it goes in only one direction—ahead. The omnipresent arrows are really only vectors on the curving, re-polarizing infinity ∞ path that oscillates between both bubbles from each cell of the membrane interface, repolarizing as it shifts between   bubbles)
The two bubbles interface in a mobic membrane that can exist

only at the mactor scale. Each mactor "pore" or cell at this interface combines attributes of a Möbius band and a Lorenz attractor, hence the name. Gravitons drum dimensionality into existence here, knitting the smooth fabric of space and time at a scale far smaller than the Planck scale where lumpy particle-waves of matter and energy emerge.

Three graviton beats in a pore create the dimensional tension of a triangle with opposing faces of 2D space and 2D time. It is a 2DD triangle. A fourth beat in that sub-quantum topology develops a polarized tetrahedron made of four such triangles. It thus becomes a 3DD tetrahedron with two volumes, inner and outer. Both volumes project far out beyond that scale, extraverting 3D space with three stress vectors of ½D time far above the mactor scale...and likewise, introverting 3D time with three stress vectors of ½D space far below it.

So many projections by so many pores of mobic membrane interface merge into the holographic Double Bubble universe extending far above and below the mactor scale. This holographic merger renders smoothly contiguous the two bubbles of 3D space and 3D time. It also consolidates the many vectors of ½D time above and those of ½D space below. They merge into each bubble's arrow of only now or only here. Those arrows are really vectors along the oscillating infinity ∞ path that repolarizes into either ½D time or ½D space at the border crossing between each 360° circuit of a bubble.

How many dimensions are there in all? With 3D space above and 3D time below, with ½D time above and ½D space below, along with those tiny 2D space x 2D time mobic bands in the pores of the membrane interface, this gives a total of 11 dimensions for the Double Bubble universe.

Thus the Double Bubble's working dynamic recalls a Lorenz attractor, for its two domains are the upper and lower bubbles. Its 3D space and 3D time act together as the three coupled, ordinary differential equations (ODEs) that allow both bubbles to iterate along the oscillating infinity ∞ path the evolving non-linear solution of emergent, ongoing reality.

If we track the arrow of time back to its origin as a vector on the endless path oscillating between the two bubbles, we find that it can indeed reveal a strange attractor—the one we live in. This proposal of a fractal Double Bubble universe could account for the continually emerging chaotic flow of events that is subject to abrupt, seemingly random changes, yet that has somehow also managed to evolve from hot plasma gas to ever-diversifying matter and energy in particles, atoms, molecules, stars, planets, people. Our universe acts deterministic in its basic fractal space and time forms, yet its matter and energy contents will iterate the ever-evolving specific details, including our lives, in its ongoing non-linear solution.

0 dimension (Quarks)

In referring to our ancestors we sometimes say we are standing on shoulders of giants. The level underneath Lamba has to do with traditions or rises of passages of our lineage. There are 8000 quarks in existence but on earth mainly 47 are respected or found here on this earth. [Quarks can be cleansed by using Purple and White Reiki energy waves in tandem]

The religious bedrock of society. To clear this level one must persevere with being a seeker of 10 religious paths as modelled by our society or parents. Each tradition has Math, Music, Physics and Magic. Its own language and customs.

1. Palaeolithic   245561 (sing missing burial song)

2. Atlantis   245617 (one omni verse clear)

3. Ancient Near East   621257 (green, brown, yellow)

4. Sumerian   561251 (gold, black, yellow x 2)

5. Hellenistic   521345 (white, black, grey)

6. Egyptian   625143 (red, lilac, black, grey)

7. Judaism   567827 (brown, red, purple)

8. Zoroastrianism   723456 (white, blue-grey, yellow)

9. African   271431 (gold, yellow, blue grey)

10. Shamanism   321547 (pink, purple, red, blue)

11. Hinduism   786245 (gold, red, white, blue)

12. Jainism   653241 (black magenta, grey)

13. Roma   652145 (red, tamarind, green)

14. Buddhism   525312 (white, red, gold, grey)

15. Zen   324567 (white, black, grey, pearl)

16. I Ching   654321 (yellow, green, blue-grey)

17. Confucianism   745623 (turquoise, yellow, red)

18. Celtic   653241 (green, yellow, black)

19. Neopaganism/Wicca 725463 (green, red, black, white)

20. Aztec   653421 (or clear every two)

21. Gnosticism   536241 (clear Cartesian plane)

22. Christianity   624351 (clear all yellow)

23. Ramakrishna   785243 (red, black, blue-green)

24. Namdhari   679851 (red, blue burgundy)

25. Theosophy   523465 (pink, red and black)

26. Science   14, 12, 10, 5 (light -royal blue. green)

27. Christian Science   423452 (royal blue, yellow, white)

28. New Thought   562412 (white, grey and gold)

29. Oahspe   652442 (red, ochre, Tabasco)

30. Age of Reason   564232 (white, dark blue -grey)

31. Existentialist   652545 (utter darkness)

32. Basque   652521 (yellow, red, black)

33. Sikhism   672717 Tupenary, purple, sage)

34. Islam   572715 (brown silver-green pink)

35. Shinto   958525 (magenta, silver, grey)

36. Bahá'í   623424 (brown, yellow, red)

37. Self-Realization   653521 (pink, blue, purple)

38. Caodaim   562421 (grey, purple, light green)

39. Rastafari   652518 (green, yellow, black)

40. The Church of Satan 242456 (all in utter darkness)

41. New Age   652645 (Purple, green, orange)

42. Native American   552534 (Turquoise, Celtic, black)

43. Icelandic   243567 (Norse, dravic Red, pink)

44. Austrailia   256751 (Aquafresh/Rainbow)

45. Unitarian   252436 (yellow, grey, white)

46. Jedi   567823 (black, blue, yellow)

47. Discworld   453213 (black, grey, white)

Imagining Negative-Dimensional Space

By the 1940s, topologists had developed a fairly thorough basic theory of topological spaces of positive dimension. Motivated by computations, and to some extent aesthetics, many topologists began searching for mathematical frameworks that extended our notion of space to allow for negative dimensions. It wasn't until the 1960s that one was constructed – the category of spectra. A spectrum is a generalization of space that allows for negative dimensions. The study of spectra, called stable homotopy theory, is a robust and elegant field.

Early on in my experimentation with psyche factors I noticed that once emptied the psyche would automatically break down id passed to the dimension above it and straight down to the one beneath. [crosstalk]

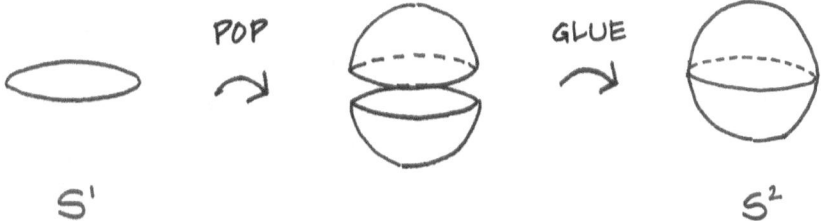

Illustration 15: Going up in dimension by "popping out and gluing."

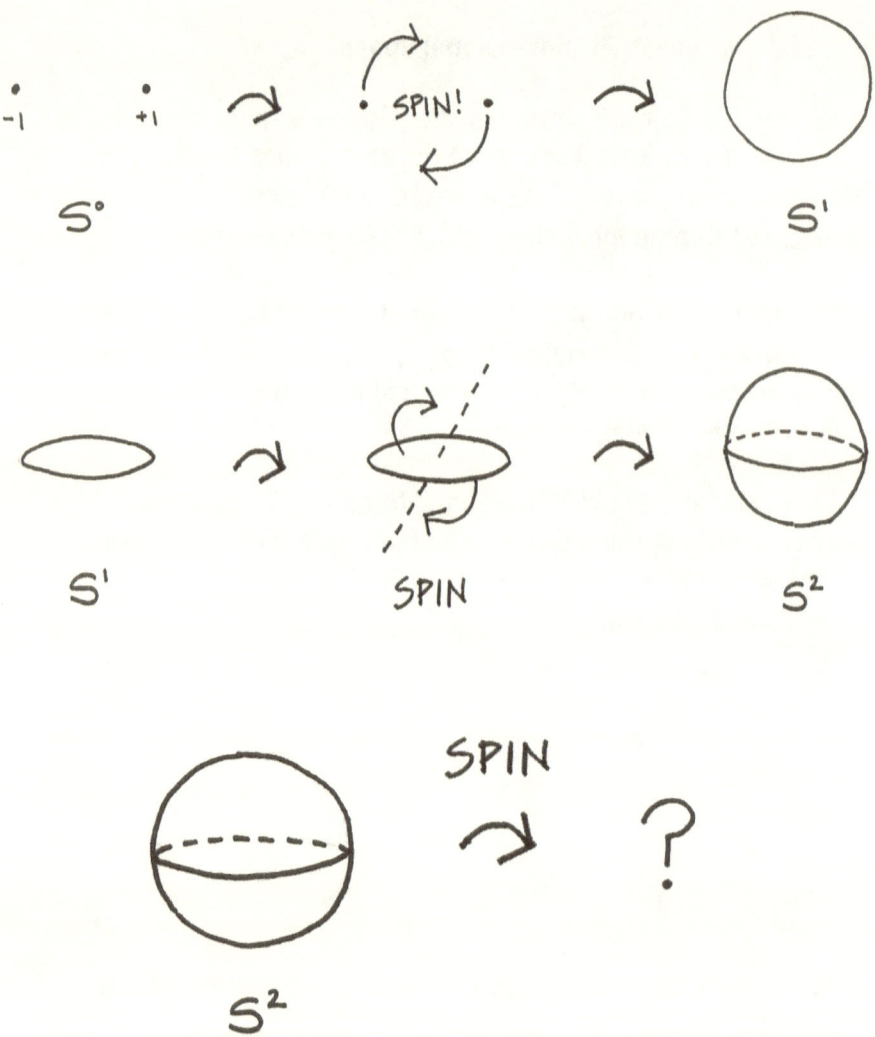

Illustration 16: Going up in dimension by "spinning."

The final illustration suggests how we might visualize moving down in dimensions, from two to one, and one to zero. What comes next? During this time you have 11 seconds to grab the psyche element per side. [crosstalk]

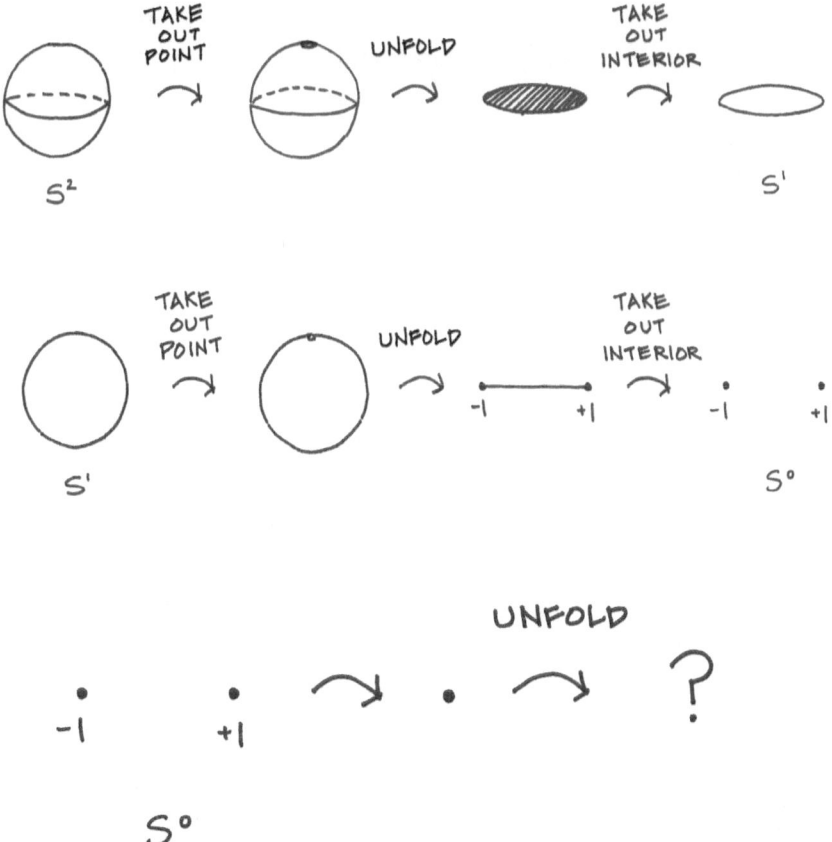

Illustration 17: Going down by "poking, unfolding, and carving."

After this action dusting, freezing, or goldening can occur. The former turns the psyche trash to dust. Freezes the geometries to the item. And goldening stabilises a certain state of wellness and consciousness of the clients psyche.

[I use White-Blue Reiki rays to cleanse the lepton scale]

171

Micro verses are dimensions formerly defined as any universe only accessible through vibrational battlement (shrinking). [citation needed] The belief that these worlds exist "within" atoms may have given rise to the stock phrase "There are worlds within worlds".

It is not actually the microverses that are microscopic in size but rather the nexuses which make them accessible. It is thus theoretically possible to enter the same micro verse from different points on Earth

-1 dimension Lepton (Soul Mates)

A leptoquark is a hypothetical particle that might have existed in the first few fractions of a second after the Big Bang. While no evidence for them has been observed, if they did exist, they would provide an important new insight into the origins of the universe.

In Plato's "Symposium," the character Aristophanes explains his theory of the origins of romantic love. In the beginning, people consisted of two heads and eight limbs. Two people were blended into one. Following an attack on the gods, Zeus split people into two and thus men and women were formed. According to him, we've been looking for our other halves ever since.

Today's Result of the Week covers a similar idea. Scientists studying the world have dug down to the very basics and found that all of the universe can be formed through a mixture of two basic particles: quarks and leptons. Quarks are found in the protons and neutrons at the center of atoms. The familiar electron orbiting the atom is a lepton. We now know of three distinct families of quarks and leptons.

The first contains the up and down quarks, as well as the electron and electron neutrino leptons. This family makes up ordinary matter. The second family contains charm and strange quarks, along with the muon and muon neutrino. The third family, the one of interest today, consists of the top and bottom quarks, along with the tau lepton and its neutrino.

Just like the men and women in Plato's "Symposium", these quarks and leptons have very different properties. But what if these two particles were decay products of a hybrid particle called a leptoquark? A leptoquark would contain the properties of both quarks and leptons, and we would find it through its decay into them.

DZero scientists looked for a so-called "third generation leptoquark," a particle that could decay into the particles of the third family. This leptoquark would decay into either a top quark and a tau lepton, or a bottom quark and a neutrino. Because the top quark is so heavy and hard to make, they searched for the second type of decay.

After studying 200,000 events with the right properties, the scientists concluded that they saw only the expected number of events, without the excess that scientists require to prove leptoquarks exist. However, this experiment ruled out a broader range of masses than earlier experiments. In short, while still possible, the leptoquark idea is now a little less likely. Nevertheless DZero scientists continue searching for the new phenomena that one day could change our view on the origin of the universe.

Sergey Uzunyan and David Hedin from Northern Illinois University were behind this interesting study. The team working on Higgs searches in the same final states also made important contributions to this analysis. [crosstalk]

-2 dimension (Stewardship)

[Clear with Green and white Reiki rays]

Dominic: Hello how are you doing. OK. So it's been about a week or two. What is going on?

Katrina: Well there's just a lot more energy unfurling.

Katrina: So I've been really tired a lot.

Katrina: Going to sleep to see things from a T. Not that I remember up when I do sessions and during the day. It's not so much during the day when you're doing sessions and you can write some of the sessions well with certain clients that can go the sessions have been outrageous the energies and just what goes on has been really strong and I just I don't know what's happening during that time but I just. Even when I haven't had sessions in one week was pretty quiet I don't think I had any. And I just felt like every day I get to session my energy had that feeling of kind of being a little bit a little bit you know tired of being burnt for other uses and things problems.

Katrina: So yes so that's kind of the overall.

Dominic: OK so I am wondering you had pretty engaging weekend it is true you were awarded yourself a solar harp so you had you have now a backup for you to make your life language work out and make targets and sensible goals for alleviating suffering on the earth from the animals perspective when they're dying when they're being called all the people dying or going up being born soothing pain Soothing Stress.

Dominic: And you've got a lot of arsenal of strings in it so the harps and then you've got a lot of potential to assist humanity which is really wonderful. It be waking up to people or people in your community there are quite a few people in Colorado that I know there are only 46 people in are fully sapient. And then the rest are two hundred fifty thousand six hundred eighty-six people who say know the causes and then they always the more people that are sending to causes that the paraplegic like the cure which stretch of your body all said and the body is a functional part of your psyche. And then you function with this as a second person and then you then you can do this and must you do it at night nudging the envelope if someone can do that during the day.

Dominic: And the device itself aren't of the right reason for that. And if what the right qualifications for that can do a great service to humanity and that's what I'm wanting to see if you'll be willing to do likewise. Do you find it funny that half of all people who are sapient and stock like an organization and I've got it right I've got the wrong idea but I want to do with it. And I've got some people over the world who are resonating with this material and this transmission wants it on their journey of being clean and sober actualised up and down and they've killed themselves up and then I need to work on with our goals and targets. That's what they will start seeing that the laws are proceeding because I was especially with my students that I've attended to taken under my wing to bring into enlightenment.

Dominic: The point is to lead to spiritual moment which will make them that if people have society in country societies once they have they're good enough and they have a good-enough-ness score to you know basically deal with mankind's problems and was in troubles and how people reach a sensible solution and peaceful times in a golden era that ensued and with the next 50 years. And so this is why I talk to this when I talk to you I've just come to somebody who a very well person you've done been the secular person with just biological clocks or being impaired or being paraplegic or being somehow challenged with motor functions.

Dominic: Many lives you just watch that. I watch that show that you've seen before he looks back to us you'll see that you have to go back in session that you want to go back with me and the sessions to see the last ones where you've had some issues you see that they sex experience in dealing with you on this level has minimal resonance level the resurgent level and seeing different things and it's wonderful to be able to show this to and to share this with you. And many people have taken through the work and have been depressed and sort of not and if they would manage anything and then switch you on to their vision and they're listening and then getting onto a point where they were the last of my global trends and what how they can help society and intuition is building.

Dominic: And I would have been an US version and some people who are celebrities and musicians who want to get got started and they had some their star struck in their eyes and that didn't start on auditions. I've gone through my I've gone through the water of life with the people and I've learned a lot on my journey about what people can handle or what people with different levels of different ages and different person different person age is just not persona and different kinds of pots that you get with your self-development. So I'm not going to get a different level here different glow you get something like every time you get something like a different aspect of the song is more mature move move is there and you're inheriting your capacity to basically assist humanity through a very turbulent part of our history this century has not been an easy one.

Dominic: And the book is going to be a curfew to channel to write down. Is he going to be getting messages and you be getting visuals and you can be getting information with different people. And I think that when we go through it at some point you do you have an education with me about how do you dream sending classes for you know moots idea. But it's funny and I've done many things with your library card just supervise you and then I'm sure you've practicals of different things and the fact of the lay-lines and everything else you learned a lot from me and that's the point of the knot of the appendage. That's the point of being.

Dominic: Being having been a student and emerging as a teacher in your own right and that's wonderful. It's I'm so happy when I get somebody to this particular stage of development and I'm I'm just as thrilled with them and I set about with them. And step out of your side. So the other good friends because of your family's not since it was hitting society the World and that that's the only thing that you cannot actually change. You have to do that and you have to. And that was actually for many years basically. And it's you know our journey together it's been a long time and it's been it's been it's done a lot of time too.

Dominic: I realize that two hours is a long time for this soldier and it's a long time. I'm an awesome person too. It's a lot of energy for one day. Many clients ended Similarly within just an hour of work with them that's quite fine. But you do like an hour or two or three hours into one hour is actually quite a lot. So if you can keep if you can if we can do sessions for one and a half hours and not try to slow two I've done that before somebody and I both. It took them a while to come back to reality. So that was quite something to do that. But like I said when you when you are good enough this has reached 100 percent we will start doing something different and hopefully you'll be able to sort of titles and happy plans in a manner. And I'm like in many schools with you before.

Dominic: Therefore away but when the students are the big change and then the big process and I took them for life for you and for a number of years or 10 years how long a days I am I'm there. I'm involved because otherwise I lose sleep overs of these sort of things that I have to manage. A few people in the end of the day. So it's quite a it's quite a lot. And it's important when it's makes it is going to happen to somebody big transformation and actually the right moment to make sure that they're happy and they then sell through it and that's that's my commitment as a therapist. That's what I made to you with that paperwork and that's what I do. It's my job so you would know that industry and you will know how it is. So when you switch on it you begin being like a married mother like a parent having them through all that all the issues and all that.

Dominic: Like Process things could be anything. Finances could be a depression. It could be emotional really mental It could be any level of functioning and it is it's a lot but I enjoy this what I find challenging compelling and rewarding. Go so I'm going to country down and regress you into a past or in the unit go through a towel whether you see it visually and hear things and feel things or maybe you will be listening and smelling things.

Dominic: I don't know what's going to happen differently but there is mean that you have a different thing with daytime visa for receiving suffering at this point. I'm not sure what I'm going to able to offer you but it will be something that it is more of a listening jury to hear the sound of the people on the earth working with the Earth maybe more of the sound scape journey of sounds and visuals, some visuals more sounds and like crazy was not the visuals and some sounds on a little bit here and the voice here were saying you are the teacher or just saying saying we're saying something was alive or doing something briefly but not a lot different from than from the rest you are getting to a certain stage of completion where this journey is very natural to

178

you it is no longer just a different animal it is second nature.

Dominic: So this is wonderful. So when I meet you in future we'll be doing other kinds of session or other kinds of work together. So yeah we've got to just close your eyes for me this might be one of our last journeys as well since I'm getting so I don't I'm not I'm not often wrong. And if you establish what you're going to be doing in future and then I can offer you any certain amount we get if we get you 100 percent you will have more than enough to give to other people.

We just find the right people. And that's a time for them to switch onto this sort of stuff if they're not. It might lead them away. So it's important to me at this point to imagine you got to teach teachers to be good for many things for many people many teachers many others. In the end you have to be the judge of when to do that to a person or not that they deserve intervention and that love actually can contemplate some something like that. So I think you been a great teacher to me and I've obviously been a great influence to you too.

Dominic: So I think that by exactly but I if I'm saying this like this in a small way it's just that our relationship is changing and the rules are changing. Can you tell it's that that you will be able to let us facilitate this was not going to be made is going to be somewhat out in eternity. Either one of the Counsellor's helping with some of your work and it is people which are visually you can get messages from them or you can listen you could do what you can to get you get admission. And then there's the it's easier. Other days of the poor and any back to the end of any you God's easy there's this much invention at that point like we initially when we support the point of interest was a couple of months ago which had most of the sort of stuff. So I'm actually OK with you have taken the journey of not taking too many journeys and it's been fine. So let's continue let's take breaths that breathe. It is the middle of an ageing century and century and you are somewhere very tribal.

Dominic: Could you tell me where you are doing things today you might have a vision that chemical's about how to do this thing but I'm finding the kind of journey where I can sense a lot a lot of kids often I don't know something isn't it's like you can like when you go and then there's like people and there's a very kind of Americans that smell after that you had in America and that of the very is something like that just like I think you are maybe a little north of there but it's an area something like much she can get the air clinician's something like that.

Or maybe even that I'm not sure I am seeing it it's a little different. They're showing some There's a sense of some kind of relic of that of these people. It's a very sacred relic with I don't know what it's up. I can't quite get a fix on it. Make something that's been passed down for generations it's really old. I don't know if it's a pipe it might be like. I don't think it's actually soapstone but it's like the soapstone. They call them peace pipes. That's a ceremonial pipe. Wow I can actually feel it in my hands. There's like a transmission that's happening in this thing it's called

Dominic: you already have a Mirror fur doing counselling work in your office in your home and you have some few items that have been augmented ...

So we dance like you did with me when you went you went and you went to the sun and then you came back you could do what you wanted to. If you get what happens. And so then he said past the present and it's kind of like read materializations. It's quite an amazing thing. What are you feeding into this instrument. Is

Katrina: it like that. Yeah just like a pipe or maybe it's a flute. OK. I can't see it really clearly but with my hands and there's energy.

It's transmitted to me. Is it actually said something about instrument because that's what it is. It's not an instrument. I can see it as it is.

Katrina: I used to I used to play the flute was only in very special occasions but it's a very very ancient instrument. Wow and it's just I just can feel it in my own field. And I played the flute. It's another instrument for harmonizing it seemed to work particularly with the water being with the water pressing on it and in the water in it not the water. There's a connection and I see somebody playing and they think it's me there's energy coming out of the waters emanating out of water and it's like the spring emanations coming out of the water.

Katrina: What is that. Wow. It's doing something for the water beings to allow them to hold certain like nurturing energies for evolution for what you or for screeching so. So it's become like they become um I mean we already are but they become more elevated in being like a nursery. Four times of rapid evolution is what it's seeming like. Yeah I just kind of like a confirmation that's exactly right. And it's speech. So it's like baby pink energy. So in school I write like an anthropology and there's a serious point evolution. Evolution doesn't take place slowly a million years.

Katrina: There's various points where it's very rapid and around a certain point it's not all that mass and it's 18 percent chance in any sense in 125 percent when the water beings come in to each other to help facilitate this very rapid evolution of course we know that you know presently for sure many realizing what I told you previously that you saw a certain conflict genetics for harmonizing and you saw that you were working as the equivalent of a blue cloak the eye of the other either one or two of the other columns and three rows. But I know that one of the ones is to watch the evolution of the twin water bear as you might be talking about the two of us was the genetic diversity and ethnicity and then a bit of genes and things on the planet with you they watching Anthony the other guys in the red and the green light on the.

Dominic: So it's a lot of information that I've spoken to back to you about these before. In the end it's true that you're getting an organized it analysing it and spreading and you're now formulating it in your own words which is wonderful and that means that you've mastered that concept. Beings that watch over us and this means that by nature the instructions like light not just the wings not just the city and not just everything that is. There's something greater behind everything and to be deeply with that mystery and to know what it is and do not know what it is. Is the Grand revolution Illusion one can have in life a mystery.

Dominic: This is where we get to the point where we were entering into mystical stages of mystics mystical many centuries. We can have this infection even as he was on the most mental level of what was a public good thing maybe only was something that did what they could have. This is an era in history where we are able to do many great things and it's wonderful to be in this era of history.

And once you get to work at once you get fed and everyone happy it's going to be wonderful planet it is going to be outrageous and be beautiful. I can't wait to see seeing this instrument and you're going rapidly into this phase of this process which is wonderful. You didn't need to tell me it's going to be getting this item. You don't need realize and do something with what it does mean what the balance with just meeting with a community and in it is being with us and you know it could be many things many that you run water once in a lifetime scene where you are dealing with much more weather woman. Yeah.

Dominic:So you. I think it means a lot to me but I don't want to I don't want to make you do anything .I want to see what you think about this thing. What do you what do you think about what is the path to present what does it mean.

Katrina: Well I think I can use yes to do this now which is kind of what I want to do and I do have to ask I do tend to work with whether even now I'm a pretty pretty attuned to it and I will sometimes ask to rain or whatever it usually does. So I'm familiar with the water. I feel like writing in the right place and I shall go out for a while and I just have kind of let it go because that's difficult for [her husband] just difficult for him and I feel like yeah the wind is calling me to be at work here.

Dominic: And now you have something that is that is going to connect you to nature to the world to whales and to the world. And deep deep under the ocean currents said stuff going on with the weather and you know what you discovered what you might explore that but it's quite a big arena. So yes I want to be everywhere.

Katrina: Now I'm hoping I get to bring them back into my office and put it in the fish-tank. Yes. As a nation I always. I've always laughed a lot. Now I to with your phone which is your life would show you greater than the nation and that's wonderful. Yeah. What do you want. What are the solutions.

Dominic: What is the source of life that it is. It's into it's an interval in everything that we have that we do. Right. We are made of many things approaching that water inside protein that has no assets whatever it is it's got some of the atoms and things. But it's true. You you made of many things in the end.

Dominic: Thank goodness. But yes it is sure what the Bose condensate is a function of water you'll be getting [her husband] is doing the DNA water journey with me which will be covered in weeks of carrying up his ancestral stuff and it's going to happen a lot emerges find his male side and find his ancestors who are struggling who are lovely and kind and his family with him. But I think going to have him a lot more than that but you will to deal with him so you know all the family to talk to about nothing is to keep this just just just to set it up this is what I'm doing with him so when he when the package arrives at your home that's it's a function of water.

Dominic: There's a new discussion a new level a new dilution of water that works in a different level and certain genetics and that's quite exciting so that the many people who have many kinds of what have you looked in journals water you will see for miles that did not say what it is. That's certainly better than it's ever studied or that it's anything it's a new phase of what has been discovered is another one I think on another level all the time naturally naturally mentally or so the time actually all on DNA on and on and on the traditions on the chemical level. So that level or the lepton level all the level below.

Katrina: And it's it's quite exciting. There are a number of interactions of water but it is actually one of the prime forces of life and it is so beautiful if you if you are someone who's more visual and you don't be able to motile out in eternity it has been for a long time. If you look in a certain city way at the turn and you get out of the sky and you get out of the universe and you go back you can still do after the sea turtles. Atlanta has kind of receded. I can show you the universe in the past I can show you all the levels of humanity and I'll show you all what's going on right up right up this red like a scientific journey if you want to do that with me at some point. Like I feel like I do for the kids to see me. And you just want to do sessions with me they are going to show them and you know new constellation of new something different candidates and it's kind of exciting.

Dominic: So what I'm going to say is that there is a lot we can still do together even though they are happening and our roles are changing and it's quite fine that you moving onto that these are the people that are going to be handling you and dealing with you for the past week in the city where if you look at the world for the past 200 years you've been seen that been dealing with these blessings and these persona's or people who live near saints or near afterlife but they didn't quite make the grade and there was that and they got robbed of the gun incident scales. They're now gone. But there was for a long time and there still are people who control people who have sentience and then charge into certain kinds of radiation of light or both concepts and cans of water shedding sounding waves. And that means that they're dealing with different kinds of elements in different kinds of radiation different scales that are out powered by a different kind of force and they're different things that form at the beginning I said and we're getting you know fire you have the word in that when you do everything except the speech in that space for certain face severe turbulence. And so getting stuck on a certain level of water. So you do everything except for movement visual this race to be less visual and auditory.

Dominic: But it's not movement at all. Then you get space that is that they're talking about the great Upanishads at the top on the sides of the six Apaches sets and the signs that are about the 6 5 or 6 grand forces and the 2 immediately weak forces in different traditions at all about it in different way the Maharabarata. They talk about that and when they see their body punish such big forces and they say out of these grand States does anyone says one college percent's missing for each of these people will have to work with teens to get it right. And that's what that's for some people that's how they see the afterlife. They see eternity and they don't see the tunnel and they don't see this tunnel. They see the tree.

Dominic: They don't see this they don't see the man or they see like climatology affected people they have different traditions which is generally applied to and is to see eternity and they there are six of them. Paul never heard of and in the Abrahamic traditions the four of them. Judaism Christianity Islam and arrestors and in that sense a lot of those things if you have any one of those you could use the division of the afterlife is different.

You're going to exhibit a scientific way and you're going to see light forces math and big math and waves and dots and things. We're going to see by physically people beings at different levels for a while or you might see each of the six seven ways of doing things in a certain way that you see this. And that's wonderful. The first thing you need to know about yourself is that when you play that he's going I'm glad you discovered this new instrument today I'm going to test and I will do something with it today.

Dominic: That's what you're going to have to know when you like me introducing people to these rum's. You're going to need to know how to deal with their different vision of what goes on inside this it inside the universe inside the super and it shells that we can hear the shells after the draw. So there's an optimism too that we are over the top of the curbes You have to know that there's a difference right in you and they're going to have to be able to spend it to them that they are interested in tradition.

Dominic: They're not in a certain tradition that has been in that family that they're inherited. Let's get back to our topic. Let's get back to the water. You have a lot of access to shallow water. You see to me that you if you've struggled for a long time the sense of physicality and with movement and you have all the senses in great measure this couldn't have said that when the people searched senses there a certain way. I have an understanding that they're going to hit in a certain direction to become unified with the part of the certain part of source observing part of eternity out there however they see it.

Dominic: They have many of the blessings of many of the worlds of the skies and in the directions of saṃsāra. the wheel of life is saṃsāra that any of us catch the wheel and that everything of everything I've thought of an action like things that are in the water with my blessings or the crime blessings or laughter blessings whatever that additions to are that different people like Zen offers laughter and tears and then a silence focused on being silent. And so you know each of these traditions actually looks like it's actually preparing you for a certain way of functioning in many senses except for one of them

Dominic: that appears to be for me the words that I'm trying to help people that the father of a person can see is out there in eternity it's very far away at it happened 800 trillion and 67 units levels away from us. It's outside of it and it hasn't this expanse is vast. So whatever person can see beyond that it will be based on their genetic capacity based on their spiritual development to see what's going on. But when I think of this I always look and see what direction they're headed ultimately. I don't use them to the ultimate topics and tell them all that beginning stuff whether it be at the end of their road with me one day. And then we work on the small stuff right down to the fact that they're alive. And we do this whole journey involved in this show them the big stuff and then I take them right down to where they actually are and where they have to work from.

Dominic: So some of the scientists approached you and shown you that this is this is possibly how you were conditioned for a long long long time. But you are living in a new era where there are fewer restrictions. You got less frictions in the way in the past.

Dominic: So whether you see it like the corner of the Cornish people or we see the tree of the gods and in Luke Chapter this for the piece they see that this tree of God from the other the gods and then the rest is restitution. They have a family tree which they see often in certain rituals and they get to see that the Hindu people for in some way also this and that this and this and Shakti and Brahman and everybody else is feeding in a tree somewhere. However you see eternity we don't have any you and your traditions or your faith. That is not something that you just believe it is. Decide what you want to believe.

Dominic: It is something that's deep inside one's person that is from the Heritage that you found has been that there's been believers that have been faithful in a certain tradition in a certain way for many many many years. You can't get that in print very easily but a certain talent that was didn't have a scene a certain way. There are certain ways of seeing in visual processing that goes along with what you get from your family which is why once you get the blessings can we get to a point where we deal with. We deal with the blessings we deal with. Getting clear on the many levels above us and we mentioned briefly there that's one element where we were three-sided three sides left that you know breadth width and height.

Dominic: Then you deal with people's development to see if they can deal with if that many futures that they've had before they must be posthumously presence is is one kind of force with each force going out. So you know that there's something like that. And you deal with that all you can do with spiritual development along growth is what you claim feel self watching most of the lessons of life. Had you said by becoming coming if you have advanced evolution from manhood to womanhood to wisdom to sapience to that which is 2001 first developmental stage of the white race for the White race. And the point is the five thousand one which is the great one.

So in this very post and the one person who's got the math and wisdom six years of development would be able to really achieve differently and the additional pages in the I-Ching it was going to be the scale of pages to be read in that book the many great spiritual books and additional pages to be discovered as you go on that journey. These are just three sides of development that I see in a special person. OK.

So back to water and back to what you were saying about the water blessing and what you were saying about the being that came to you and the faith and how you feel about this instrument. We've got we've got an hour and a half to expose and we have a lot of time to apply practical to like an oil spill some disaster some some just something with numbers of depleting numbers in the 70 population population genetics. It's going to work out together which is quite fine. Doing it with you is going to be exciting and fun is give me the history for that you need to kind of a background before you start giving these bones to many people. I know you're going to start doing this. I want to prepare you for all of that.

Katrina: So you want to come back at this time because you can listen to my speeches and my what I said on different levels. If you if you if you if you get back to the sessions and you listen to them again and this is then when are you going to give you you session tapes and then you can have what have you tempted to do yourself or you have my vision and my notes and whatever else it's quite trying.

Dominic: Describe where you've been taken when you were mentioning about you know working with the plants that are starving and still leaking radiation. I understand radiation and I don't know why the motor of the motor show that they were able to send certain water with certain kinds of energy. I mean not just these qualities but themselves in crystals they act to

189

shiny matter that they were there were other two water blessings like blessings or whatever they want to call them shadings what they want to call them tidings or whatever in any tradition of different things but they use the channel and as you said in kind of radiation to counter the radiation inside that and then find those voters and they did they big went for many. So my thinking is that there's something to do with water. Like you said and you're going to build a project with me over the next two weeks.

They've given you this instrument that given you are still happy to have used the how you've done the kind of job for me you did went through and you have you with all these guys and you go back to the present day went backwards and forwards and it was fine. The take of that cause.

So you're doing well but this summer will you be having someone else develop and there's something that you develop within yourself. So it's something that probably they to to use this food to do something with water. So take me where you take me it might be inside my eternity my insides would be on the pad and my be any way or don't mine. I can I can drag you around.

I've not even asked you to take it anywhere you need to take me to take me to conscious going to God where your intuition is guiding you to take. And I think this for even Hiroshima . I don't mind. We've got the visa and I can go there if you need to see. Map of the earth or the heart of the earth you can go to any way it's free for all you want to be doing is going to be the you don't have to be busy to see the Earth and see the map and see see places. You can go there and you can do your own work and conscious driving and you can do your own work towards restitution. So take me to take me to Perth. If that's coming up for you.

Katrina: I would like to take you on a journey and see what's going on the present day is a new beginning. It's not always accurate and it is only that you see of my '97 it's not that it's ancient. I didn't get an accurate news at this point in history. To a certain point but that's me. And we can talk about that in a session sometime. Well so if you're up and you're getting it because you're dying then you can see that the microbes in the water you know it's polluted you know stagnant and if we go there right now we can see what's going on. Well what are some deep breaths of water. [crosstalk]

And that's what's really calling me a hard time but my focus as I find it to water is what I'm really sinking into and I know these deep reservoirs are deep. And that's really what's going in it would be fun to work with those waters because their impact will be significant. The kind of like what determines if you're familiar with water domes So their ability to transmit is pretty huge. And I can feel like five of them. So they're reservoirs but it runs deep in the earth and a colony of the water is different. Tell me what the faculty you're seeing like festoons frozen actually. So what is what is going on. Yes. Yeah I mean I am taking so I guess it looks like OK I well is just such a different quality. It reminds me when I went swimming and nowadays there are these screens that are in Mexico and it reminds me of the silkiness.

Katrina: And there's again a shoeing quality in a peaceful list and a potency in the water. That's quite different than other waters that I've ever felt wonderful. It's going to be working with the moon because the moon has influence on the tides and water and cycles people.

Katrina: So I want you to take to Pluto I'd like you to just just you to be on the food and just see what you can do with it and see if that is enough information once you're ready and once you move it gives you that kind of a nod. And you know to be ready to implement this in this way it will be watched with all these things you would think is increasing.

Dominic: By the way I'm reading speed is increasing you know. Right. Do you have an estimation and knowledge is increasing drastically. I can't feel that things are really really fast going all at once and it's amazing. So expensive too much understanding and all volumes which is wonderful. Yeah I can tell. I how I like to start getting these energies. OK you sing Pink water any see any colour if the goal is not.

Dominic: It has to be green to be good or goal and be perfect. Wow. So she was the colours around settle on the place and then let's work with the frequencies then let's try to sing along with you. Are you for this sense for the water using the flute which is blocks which blocks gently blocks no resonance voices and subatomic frequencies. However I don't know what I'm going to get on the grounds that it's contained but it's high treble that that really gets bacteria and viruses I'm not sure what you're working on what level.

Dominic: But as soon as you get the product and you get to that second phase I know that you will see something that's going to lead you to the next goal. So just follow intuition and just keep following it try this out one in particular is calling I'll go there.

Katrina: I guess that's why we're going there. I feel good. Yes. Like I'm nervous. You might not know the men from out of the ground which stated that I think we in America I know we are under the brunt of some step and it's not a problem.

Dominic: So basically we can go around on different continents different kinds of different solutions. But I want to get the signal because it's going to be very boring. I want to see you experience that exponential vision to see if I can validate what I know from stats and from the from lab tests and things to look on the water we go looking for.

Dominic: Pink everything which is is not green.

Katrina: And then you want to make it safe and you want to then play your flute just playing stern and similar to the way it does. Or my chin infarct just starts in the heart. There's that there's been some kind of some kind of membrane at it's kind of containing this energy.

Katrina: It's like it wasn't some kind of way which is odd but that's what I'm feeling.

Katrina: And it's now opening up contained. I've been able to emit what it's capable of making then it's now opening up like a blossoming flower. And then there's opalescent kind of rainbow coming the Mini Me meeting from it always reminds me of Angel or that stone I'm glad it's reaching the surface. This mission is reaching the surface. I can see it really starting to come foreword and it is it is a place now. I thought it was. I do it still under Georgia and Florida. There's a big aquifer there.

Katrina: And there's some disharmony that's being really steep from the kind of like I don't I was like like big bubbles coming up being released. And so something just bubbled up and then it's spreading over the land and the waters on the surface after the planet and it's particularly connecting in a lot with trees in the and with some sessions with clients banyan trees and places I even thought I saw some banyan trees in Florida which seems unlikely but I know they're there. But anyways connecting with some I think it's Cypress. There are Cyprus's in the top really connecting strongly to. I can see some of the animals have in the swamps like I project where we see the alligators. There are crocodile tears but I'm not finding them yet.

Katrina: And this is good whatever the energies are really

affecting. And I can see the alligators central nervous systems just really lit up. And yes it's kind of a silver gold.

It's more gold. Yeah I really connect with this with the Bator so I can I'm just looking in the eye right now. You know the reptilians usually do call me actually the birds which you know they're reptilians. Yeah. It's really affecting them. I'm back to that crystal of silica. We have a lot of silica.

Dominic: You know our light has silica superconducting. It is this is the crystal that it's the silica crystal that is moving to a different level of super conduction and it's connecting with something else something. And what does that so there's a different level of development that is occurring it seems like it's being formed from I'm going to say it so that those words are awkward. So I don't really know what that means. So evolution so like why are we doing extra interesting you know because it's a thing of carbon based life forms.

Dominic: We are made out of trials of hydrocarbons and coupling long bonding chains.

Katrina: . It's a long it's it's actually quite exciting. Coming from you can't be rude in anything that talks it can be if you show it to the cat because like antitoxin in the body you can be seen shifting frequencies of silver through the water by some energy that's not on the plant I can fix on that. But I feel it within my own body I've been taking this. I get it from Switzerland it's this really potent silica in it.

I'm doing it to help BP and the superconductor that I'm supposed to be. And I think it's working.

Katrina: And I'm just kind of really take this molecule I don't know why would I do any kind of feeling the cypress trees doing what is going on. Let's see. I'm feeling a lot of this in Florida southern Georgia Mississippi Alabama Arkansas maybe a little bit into Texas right kind of in the Gulf Coast area. That's all Gulf Coast. I visited some of these springs that's part of this water in October. I can feel the manatees because I saw manatees when I was there kayaking. There they are. So this is a detox that's happening.

Katrina: There was an oil spill in the Gulf.

Katrina: I don't think where this is freshwater still something is opening up in the cypress trees. It's like another layer has opened up in them and they're emitting energies that are connecting to other Cyprus's. I'm not an expert in Cyprus. I know there is. Well I don't know much about them except the Japanese. They grow and swamps their superstar connections that are happening. So they're like satellite most Cheeser like trees are like and the Cyprus's have connected to star writings. And something like this with the family earlier in the week.

Katrina: And they're connecting with those energies of what happened earlier.

Dominic: It sounds like you want a big three water into the Cuban land which is the cooperation of the Chinese and their ability to stop and they're going to do a lot. It's very exciting what I'm hearing from the banyan tree and I like Southeast Asia. Off to Asia.

Katrina: I know why I kept any trees. There is a tree somewhere in southern Florida where transplants are there anyway the cypress connected to that work and that work went global. It's kind of more the southern hemisphere obviously because it's spanning trees that's chopped Cypress connecting. So there's kind of a little mixing of those energies because it brings the energies and the cypress spring and complimentary but different frequency to the water.

195

Katrina: Wow. Fish and sea life. That's a good sign that the water is is doing well and not just the face and realize this is such an unending many artistic leanings people and the things I was using my sentence and my spiritual development to shift invisible structures behind lie to fix things that you've done with tumors and leukaemia and other things you should be to fix things if you do things and things. That's how it should be. There's no reason why we can't the good nature in life in a larger sense in a larger scale as being affected by people's big memory. How many people to do likewise.

Katrina: And that's that's actually very beautiful. What is the result of four years with an old group and groups in that area of this of the states. And it's supposed to produce something like the time that this is like six years and something and a days since they started doing is clearing work now with time on you all is that takes on the picture. Is. Let me see if I can find any way I like it. Six years 7 days.

Dominic: That feels right.

Katrina: It looks scary for me too. So that's fine. What you that. OK I'm switching gears a little bit.

And I have been to Serpent Mound in Ohio and all of a sudden I'm like really feeling it. And I just read some cool stuff about circumambient ops and it just kind of popped into it just came forward.

So that's definitely I mean that's high ground. It's not there's not one of them that I know of that Serpent Mound is kind of like a battery of energy. Tree-beard I want as a TV watch to participate.

Dominic: That's my strength with you on a continent like this a tree for science is true for every tradition on the earth in this sitting there and they go and they're guarded and it's quite fun to look through the trees the tradition these days the modern sense they use the phone and whatever else. But in the old days there were trees in the trees. They are ancient and that are the guardians for that particular continent.

Katrina: Yeah. It's the trees actually and the guardian spirit the energies that are that are stuck here. I don't know what kind of cheetah's it's rather temperate zone kind of tree. I want to see say. I don't think that's right now and I'm pretty sure it's not. But anyway I can really feel this tree these tree guardians I'm not sure why they call me but I'm going there. They're asking me to tap into this energy that it's like took about as a battery of energy and I have been there before twice I don't just want to in the colours of the water. What are you seeing in the colours of the water as it changes in the trees. Oh the.

Katrina: Started off kind of green and then it goes again into the stack ready to pass down all silvery white with you know pastel rainbows. It goes to the. It went red like a red six different and it's gone back to. But the quality of the screen is different to different it's a different clear queer kind of green phosphorescent green. Now it's going kind of silver into gold. It really went gold. Kind of bubbles of gold because it's cold then it goes back to sprinkle.

Katrina: It kind of going back and forth sort of flickering and now the cold is spreading to whatever kind of tree this is it's spreading over more of the temperate the northern US made middle to Northern US or whatever tree this is actually might be you know I actually might be. There's many kinds you know shocked seeing some of these Forests kind of lit up. And I think it's the Silica guy. So it didn't you know siliceous entries right.

197

So a car molecule that's doing this again is going yeah. Gold gold and silver exposes Kelly gold the silver kind of meeting in a way you like wavy lines on the line sending strong getting one more strong and it's like there's like this internet between the trees I mean it's always kind of been there but it's a much stronger. So it's like the electrical circuitry is just gotten really strong. And then it's affecting the the ecosystem of the what the trees are in. So it's effect the ecosystems and I've seen like purging in ecosystems.

Katrina: It's like they're starting to cause you know the energy is kind of big it's clearing it's clearing whole ecosystems as other trees are kind of being affected like some of the ash trees and they're having a hard time right now e.t.c So things have gotten very bright across the into Canada the US. To the Not quite south but three-quarters of the country and there's a lot of clearing going on. So there's a lot of effluent that I'm seeing across all of three-quarters of the U.S. and into Canada.

Oh wow. It just kind of went on for something there's a great release happening. However there are a lot of North America haven't seen it go off North America and Hawaii. It's almost bilious this energy that's releasing its kind of I think a lot of it is eventually it has to do with humanity the energy that still uses is it is the energies of humans. A lot of it and it is it's just kind of bilious energy. I'm on something just kind of a shifted. It's just kind of it applied to where it just went.

Katrina: Not really sure what that was. There's a realignment happening really quickly it's kind of weird and linear with the energies of the trees particularly like all these lines are re-orienting. It's hard to explain almost like a career that was there is it but it's not a very fluid created with all kind of linear and jerky jerky and it is spring out so it's not so crunchy and so to work instead of sharp right angles and such it's smoothing out to be more like a wave instead of jagged.

Katrina: And now it's enter the actual land itself.

Katrina: I think it already had I just don't see it very clearly. Now I'm just spry back to Serpent Mound again. Something like is you cycling to the response that always really emanated from serpent. When I started and then I came back to me and there was a whole country as a whole. Yes.

Katrina: Yeah. Yeah. Oh was set up. And she was just like we just kind of reboot it.

Katrina: We rebooted the great work circuit mounted. Yeah I know it feels like the circuitry is clear and open to a greater extent and it was just kind of a reboot. Sappy things feel kind of dumb for the moment. Now it's interesting. Now I'm just seeing kind of gentle emanations from the land but it but the event is over and there's just kind of like pre-increment land just a gentle like ah that feels better. Kind of a sensation. I don't know that I use my twit for that but whenever.

So now I'm feeling that this will be interesting because now I'm feeling some of the snow days in Mexico. I've been in some of them one of them and there's an extensive system of underwater caverns in Mexico and then they do eventually make it out into the ocean. Miles to miles of underground water system. And when I connect there I really should tell them my people. I really feel this energy is it and I connect things I start becoming aware of all this like mathematical equations and and top two courses.

And there's just it's like a schematic a mathematical schematic is there. So this is in the Yucatán area you can across kind of chip quantum all that that the whole area where the fire works. Wow what is the if this be this energy schematic that I see is connected to the pyramids and that it's I really can feel a couple of the pyramids. I think one of them is anti-cop. I haven't been to the call but I can feel it. Another one in Conjuring a.

Katrina: Then there's the big pyramid of Chechen Itza. I really know this area and then there's another temple a smaller temple. I don't know exactly where it is but it's it's a Jaguar temple I know this temple. I didn't hear before come to play my flute at this Jaguar temple is what I just heard. This whole thing kind of reminds me of the Egyptians and how they have different temples along the Nile and it was the pilgrims went from the first soccer temple the second chakra and the you know over my ex worked their way along the Nile along these temples and there were various teachings and mystery schools associated with each temple.

This reminds me of that its configuration is very different though. Anyway so here goes. I'm getting ready to play the miracle. I feel very connected into some star energies and that the these are just coming through me and I'm just kind of channelling it. I use it you know I was in the now just felt a lot of energy just comes through the temple and you have a temple kind of activated.

Katrina: I think I did go out. I saw the sun disappeared. And do you do if you do the pleasing no I feel more energized to be. Yeah I had a lot in my head.

Katrina: It's like I'm just cutting into circuitry here. I thought it was going to be for the benefit of it but prong for the benefit of the wrong direction. I. I'm not seeing anything I just feel a lot in my head. It's my central nervous system is taking on. I want to say it's like a mile. There's a sheep that's just coming to my service my central nervous system is how it feels.

Katrina: You know like balm of Gilead or something. I'm just feeling coded in a good way. Like the skits are being created it feels so good. My. I'm good.

I guess you're getting called in the order is something that you are meant to you in a certain sense a part of you develop and a part of your nature that can be synchronized among all your developmental parts. Not only do you have a soul and often a presence my net and many things you have many of the many many different kinds of self-development. Nice developmental motifs and possibly some transactional Ponson the transaction consonances closed again. So that's when you remember that talk about possumus was and it wasn't just by the United States but the idea was broken down or anything else. Some are those guys wasn't like this. To me it was about all it was different. And it's that you had different parts of it spectrum of human pain in your eyes to the right to your vision.

Where did this one gives you this gift and that gift that sense or something different. Once you lose that happiness it is all you will all problems. You have to have something to function with. And so then you develop more parts of yourself that it can extend itself. It's like if you remove a child on the stand and you see some incident referencing some reading of that book and you have enough to kind of see what's going on. This is something that I developed when I was I was in this I was sort of reclaiming something that I worked so good.

Katrina: One of the things that's been happening since this last 10 days or so since I last session. Is that big increase the amount of energy coming through or just on a regular basis. And then my nerve endings feeling like they're burnt. I'm just feeling kind of exhausted. And that kind of playing out of my relationship you have some difficulties I just don't want to do anything. And this just feels like a coating on some of my nerve endings in my whole nervous system that feels really good. And it's started in my head but it's just kind of got to my whole it's going it's not quite done it's going through my system.

Katrina: I get like I'm three-quarters of the way through. But still a quarter to go. There's something about me connecting into this grid work that allowed access to whatever this is trying to reconnect to it. And there's a really strong connection to Venus a lot of that a lot of the temples are connected to the US in Mexico a lot.

Dominic: That's interesting.

Katrina: Yeah. You say you know from there that no starts connecting somehow to South America. Say that again the night star Venus is connecting to with your friends to South America right now in Mexico. Mexicans the people around there connecting with the plants and animals. Yeah yeah. Well I think actually South-American there are temples to business there too. But I know that that's what I'm feeling and think I connected to the next generation said I know and there's there's a temple on Venus that I'm seeing I think this is where the temples in Mexico probably went to South America too but I'm connected to temple I'm some beings too. Hello. So the temple is very it's it's not it's a cool stone like it is in Mexico. This temple is quiet. There's a council here. They are above the meme (units of culture) level on the 14th dimension. Mascot(s) and Masters.

Dominic: I know them. I know them now, I know them later.

Katrina: I've seen them before. Anything that you know if you see it from a different angle that you're going through what you're going through you coming back. I wasn't back there. Coming back back to the temple that temple. Yeah. Yeah.

Dominic: The temple is absolutely stunning. [crosstalk]

Katrina: The emerald-green like crystal green and well then there's it's emerald-green but it's holding gold silver light again. But the temple itself is just green green green on the top of it is this jokester stream of energy kind of gold with silver in the center of it. And it's a live with the green. So I didn't see any of that prior. I'm here I'm really feeling the orchids in the room. It's in some of their fern right guys. So really actually feeling those kind of plants and feel the temple they're not shirts they're these. I don't know.

I'm not sure what kind of plant they are but our ancient ancient ancient big leaves cut their leaves very solid like that a rubber plant anyway just really feeling those energies. And they said sort of like this crystal lattice-work in the jungle. This crystal. Latticework So Crystal in white moves pictures of the crystal water molecules. I mean the crystalline perfectionist. It's stunning. Not just on my familiar little tank it'll take three hours to show everything all the all the stages of growth but it's not what her mother was doing with Mrs. Norton. He was highly sought after by many people and eventually many people he was knighted and then he was got another public fame for this. And then unfortunately he passed away and under mysterious circumstances so no one will know really what the full story is of the mist behind water.

Katrina: I again realize that the team of past so I've seen these crystalline structures is very is like Emoto shots of the really beautiful crystals and they're all connecting in these kind of a lot of the six-sided there's six sides to most of these crystals where they connect to each other. It's it's a it's kind of like a crystal and latticework that reminds me of honeycomb because that's you know the x-axis Conair to T-Rex's and I'm hearing that that's the expression of the mathematics that I receive. This is the expression of a is before all these formulas and I saw this girl at work and it was very diagrammatic and this is just this is not this is just this is a crystalline structure that's just absolutely stunning with the perfection of it. It's like the highest level green fiberoptics It's just amazing.

And it does go all over the jungle it's just in a certain area. I kept expecting expecting it to spread but it really hasn't. It's just continued to I just continue to see it sort of deepen it into a point of like ice as it gets you know ice freezes deeper and deeper into the water that's kind of what was happening with these crystals. Right. There was energy that was emitted and it's going to be not OK. So it's just kind of going to handle zooming into the water and seeing the molecules things we all can see the planets and so on. That's quite exciting with these two were full craters so you see that.

Dominic: And it's wonderful that you've chosen to do that you're on and the one with you doing it to the country that you're in.

Katrina: And wonderful that you understand that you're going to present you know you've been in before you've been to approach these past before so that you can remember and that's wonderful. Actually. So I can get back to my ceremonies and did work and it gave me quite a lot in my process. So just giving back some of the beam of energy from this crystalline structure to Venus is just pretty solid pretty solid signal back and forth. So it just connected up essentially to that temple and to this kind of council whoever they are that I actually do. Now it now the temple on Venus I just see it just like the eye of horror.

Katrina: I just see this. I am very strong over the top of the temple. I didn't see that before. And I really can feel it in my own third eye too. Reminds me like of a peacock. I mean it's you know that iridescent Peacock male peacocks other really reminds me of the summer time you heading out of the fitting It's it's time to get you back to your body. You zoomed in and zoom out he did many activities and it was just wonderful to watch and to witness this scruff with you in this particular day with you to spend some time with you next week.

Dominic: We're about five minutes to wrap up and then we'll get to them. That I'm you will have to do few other things. OK. Five minutes.

Dominic: I mean that lead into your body can we go who do you give me permission to put you into your body when you can. I can see that lovely It doesn't take long our wanting to see you want to meet me on Monday or Tuesday.

Katrina: Tuesday would be great.

Do you want to do it. I can sure. So that's. Time was 10 a.m. was that right.

Katrina: Yeah.

Dominic: Thank you very much. She's done a lot of work together today and it was wonderful it was very productive and it was too. I spoke to some different places and it was really wonderful to see the links between everything we have.

Dominic: And next time we meet we can continue with this project but even with different licenses to do projects around the planet which is wonderful and an ongoing thing that combined we combined arsenal of instruments and things to do things. The first thing is the residents of that is very deep it's very much to the good to do with the water. It's. Frequencies they the of this stuff that even the plants that recognize stuff in the minerals and the water. So that's what we've been learning how to really go back and listen to this again you'll see many things about yourself what you've commented on what I've commented on when I was commented on and it's a very beautiful session.

Katrina: I'm very pleased to have done this journey with you and as it was so pretty. So what is going on in the world with water and that's wonderful. I want to thank you so very much of you.

You just continue to flex your body and do what you can what you've learned to do to climb back in and do what you can. And then we will see each other again next week.

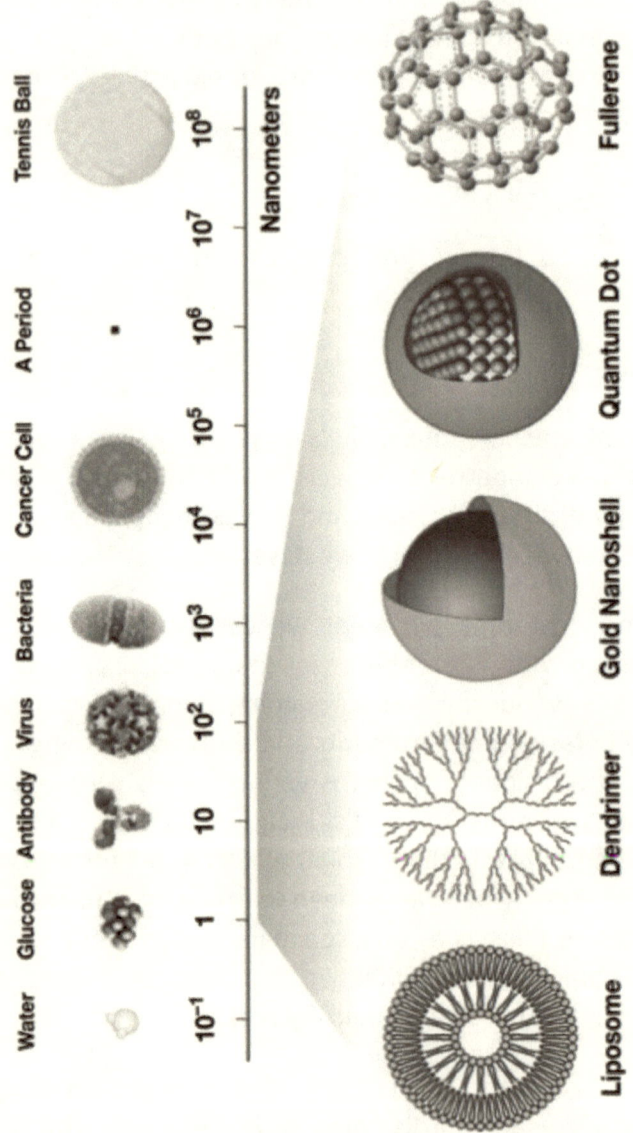

Illustration 18: Molecular to gross scale

-3 dimension (Nano-scale)

In early 2017, I was waiting to see if I'd be able to go to work after the death of my mentors in 2016. I received a notification from my Shaman Elder Maggie Whals.

My studies in Reiki had been unlocked further. It was around 10 years since the first award. It had grown to include a sound understanding of physics, music, math, healing, agriculture diplomacy and faith.

Subject to the usual ethics checks I was overjoyed to learn that I could paint kanji characters for my clients benefit without Elder Maggie's supervision. Here is a sample of a teaching that cleanses the femto scale:

-4 dimension (Legal level)

[[To be treated by mixing Blue, Red, Yellow and white Reiki rays.]

So Michelle contacted me about problems in her family and places where she needed legal support in ensuing court battles. She was worried that her husband, a lawyer in his retirement, and getting taken advantage of at work by an abusive son-in-law.

Dominic: Good morning to you, Michele.

Michelle: Good morning. Oh, let's see, I don't have the camera on. Do I have it off?

Dominic: Okay, so last week we worked with ... Um, the chakras, and we cleared a number of them.

Michelle: Mm-hmm (affirmative).

Dominic: And we also explored ... Some fragmentary, past life memories ...

Dominic: We talked a lot about shielding, we talked about the family. And we talked about how there was some anger, and you wanted to know how to rise above that and how to process all of those feelings. They're part of this journey, so recycling energy of hope, love, and the EFT. And you could even do the trauma release exercises which I sent you. The trauma release exercises to shake stress and tension away from your body.

Michelle: Okay.

Dominic: Um, but the EFT in general is one of the tools you can use to recollect the emotions. We talked about shielding and you wrote down some symbols, you wrote down some sacred geometry, which I gave to you and you got the impression in your mind's eye.

Michelle: Mm-hmm (affirmative).

Dominic: What was it like, the last week, the last three days after last week's session? What was it like emotionally processing all those things?

Michelle: I, I'm having a hard time remembering it now. I should have written it down. Usually it's intense, but, um ... It was kind of a crazy week.

Dominic: Mm-hmm (affirmative).

Michelle: But ... It was, it was my husband's week.

Dominic: Mm-hmm (affirmative).

Michelle: He was kind of, um, traumatised. (laughs) But by the end all I know is I called, or my daughter called me about my older daughter and we had a great discussion, and went over misunderstandings.

Dominic: Mm-hmm (affirmative).

Michelle: She said, "Mom, why didn't you call me and ask?" I apologised about ...

Michelle: ... You know, not asking right away 'cause I said, "I don't want to bother you." And she said, "I'd rather you bother me than have a misconception."

Dominic: Mm-hmm (affirmative).

Michelle:	Yeah, it was very good. Very good. And I found I really, I really liked this. I kept saying, um ... Like, by the end of the week I was more peaceful.
Dominic:	Mm-hmm (affirmative).
Michelle:	Because I had been, for a couple of months, taking some sleeping thing, like Lunesta, something.
Michelle:	And my, when I went to my integrative physician, it's kind of they mix, you know, holistic, natural things with ...
Dominic:	Mm-hmm (affirmative).
Michelle:	She was like, "Just don't go there." You know? "You don't need to do ..." And I, for the whole week, didn't take it and I'm perfectly fine. I could sleep fine. (laughs) I didn't need it the first night, it was four hours, second night six, and ...
Dominic:	Your sleeping fight was better as well. Lovely.
Michelle:	Yes. And, um, yeah. And I just had this feeling of I choose peace and love, I just can't ...
Dominic:	Wonderful.
Michelle:	I physically can't stand that negative feeling.
Dominic:	Yeah, so you've gotten to a point where you're getting to the white in your mind's eye, and you're getting to the object of peaceful, serene, and calm.
Michelle:	Right.

Dominic: And you're getting used to shifting into the state to live there, basically, the whole time.

Michelle: Right. Because ... Yeah. I just kept saying, "I choose peace and love." Because it feels so bad to be in the other place.

Michelle: And I realise it's a choice.

Dominic: Absolutely. I think for two weeks now we've been talking about how to shield one's self and how to kind of lift one's self out of negativity.

Dominic: And to peace and bliss. Peace, peace is quite a high vibration, bliss is a little bit more. But it changes.

Michelle: Right.

Dominic: With the hope, love you recycle the negative feelings from shame and guilt to anger, from anger to hope, to hope, joy, joy to hope and hope to peace and peace to bliss. At some point.

Dominic: Your emotional profile and your frequency lifts. And you can see that your futures that you attract to yourself, what you manifest, you manifest better things because your vibration is, is better, your frequency is better, your ... The number of futures decreases into-

Michelle: Yeah, I've been there before.

Dominic: Yeah, you were in your bliss and you spoke a lot, we spoke last time about when you were doing *The Course in Miracles* and the angel came in ...

And then at that point we wanted to do your listening a little bit. And we readjusted, we opened up the listening in one ear to one level, the lower astral, we opened up the other ear to the mid-astral. Have you had any, heard any music or any spiritual messages this week or not just yet?

Michelle: No, not just yet. So it's interesting because, probably in the past year, it's happened to me a couple of times where I would wake up and hear music that I couldn't quite ... It wasn't like on this level ...

Dominic: Mm-hmm (affirmative).

Michelle: Like violins or something, but I couldn't quite hear it. And one night I said to my husband, I woke him up and said, "Do you hear that?" And he heard it too. So I said, "I'm not, I'm not imagining ..." He said, "No, I hear." It, it was almost like a radio ...

Dominic: Mm-hmm (affirmative).

Michelle: Like an old-fashioned radio where the, it wasn't quite coming in clear. Do you know what I'm saying?

Dominic: Mm-hmm (affirmative).

Michelle: Like, like in, on this level. But we were hearing it. Do you know?

Dominic: Hm.

Michelle: Like just not as clear as if it was on earth I guess I'd say.

Dominic: [crosstalk] spiritual listening is a lot like that in the beginning because when you get your listening levels in the beginning you start to hear, um, the vibration, you feel the vibration. Then you hear it softly, vaguely, like a whisper. And the, the sound gets stronger and stronger, and clearer and clearer and clearer.

Michelle: Huh.

Dominic: Okay, so how about this week we do a little bit more work with your chakras and then we start working on, on, on ... Of building up your listening, your clairvoyance?

Michelle: Okay.

Dominic: For the listening that we want to get at today is the listening about the past, is about the listening for the causal level so that you can hear it inside. When you, we see the scenes of the past lifetimes in your mind's eye we can, you can start to hear the sounds of the environment and what they were singing. So we see the Navajo, like you saw one time, you saw the vision of a, the Navajo healer and you were a midwife giving, helping your, one of your daughters give birth, and you could then hear some of the language, some of the sounds. If we open it up on that level. Okay, good. So if you can close your eyes. (call drops)

Okay, we're just going to keep the video off for now. If you can close your eyes and ...

Michelle: Oh.

Dominic: Go into a space of white in your mind's eye. Can you see the white expanding in your mind's eye?

Michelle: Mm-hmm (affirmative). Yes.

Dominic: Okay, now I'm going to bring you in, bring you a mirror, a mental mirror into your awareness in your third eye. Can you see the golden bordered oval in your mind's eye?

Michelle: Yes.

Dominic: Okay. Now, we're going to work with some of your chakras. You could start seeing patterns and petals and chakras and things, or lines and energies and colours. Let me know what you see. You might even see some pictures of the chakras, opening of the petals, and then you might see some pictures of ... Like last, when your base chakra you had pictures of, um ... The time before last you had pictures of where the, and you had this Native American life that you were working with rain songs and things, and you had flowers in your heart, and different kinds of images came up. If this happens again just let me know and we can dive into it, into the past lifetime and see which things, why it's like that. And that could be a very lovely thing to, um, explore.

Okay. So we're going to work with some of your chakras. We're going to work with the back, your eye back chakra. Just checking to see which chakras are being fully enlightened. Your crown chakra needs some work. Your back heart chakra, your crown chakra ... Let's look at all of them and then let's, we'll see what, um ... Okay, so we'll look at your, we'll start with your base chakra and we'll work our way up, and we'll see what else needs to be done. Here's the visual: nothing in the front, nothing in the back. Moving up to the sacral centre. 210 in the front layers, in the front ... 390 at the back. Then there is, um, your solar plexus chakra. Almost there.

(call drops) Okay. (call reconnecting)

Okay. So ... We'll be working with your sacral centre today, and then we're going to check the others as we get there. Okay, we're, here ... Okay, we'll maybe check them through first and then we'll go and do the, um, enlightenment of all of them. Here you go to the solar plexus chakra. There's quite a bit in the front and the back. The heart chakra [inaudible 00:13:04] the front. A little bit in the back, not much. Throat is mostly clear, at the front and at the back. [crosstalk]

Okay, and so we're going to work with the crown and those chakras that I've mentioned. So here we go, and let's, we're going to work with the sacral centre. So here we go, if you could focus on the sacral centre. Can you see the orange petals? Or the orange colour-

Michelle: Yes.

Dominic: To fix the petals, we used different love frequencies to fix the petals. And I'm going to assist you to open the levels, here it goes. 100,

200, 300, 400, 500. And you can see it starting, it's, it's now ... Becoming enlightened, it's become illuminated. [crosstalk]

Michelle: (Skype ringing) Hello?

Dominic: Yes.

Michelle: Sorry, I think my computer ... It's a ... I lost, um, connectivity.

Dominic: Okay.

Michelle: I think it was my computer, I dialled back on. Okay?

Dominic: Great.

Michelle: So now I can get back into our activity. (laughs)

Dominic: Back to your meditation. And we're looking at your sacral, your centre, your sacral centre is now open. Can you see it's now golden?

Michelle: Mm-hmm (affirmative), yes.

Dominic: Okay. Looking at your back sacral centre it's now orange. And a little bit golden. So we started working the petals back. 100, 200, 300, and that's it, it's all golden as well. So you can bring up the, we can bring up the Kundalini. We can bring up the Kundalini energy. You're bringing up the Kundalini energy and, um, with that bring up the Kundalini energy you can see it pierces through the chakra, the centre of the chakra and expands and flows out of the chakra. We're now going to work with your solar plexus chakra. We're going to work with your solar plexus chakra, here it is.

Okay, so we're going to work with your Manipura chakra, the solar plexus chakra. Seat of the self esteem. We can see it's yellow. Open, open leaves, yellow, yellow petals. We have got 2,000-ish petals to grow through here, it's about halfway complete. As we do this we are saving some legal repeats and dialogue to the legal level and earth star chakra level for you to build your legal case. So here we go. We're going to now open up about 2,000 petals, 3,000 petals. 100, 200, 300, 400, 500, 600, 700, 800, 900, 1,000. 200, 300, 400, 500, 600, 700, 800, 900, 1,000. 100, 200, 300, 400, 500, 600, 700, 800, 900, 1,000. And that's about all we're going to do today. You can see it's opening up. We've fixed the chakras that we needed to and you can see the fire, the Kundalini has a fire in it, illuminating that centre. Now we move to the back solar plexus, with permission. And we start working here. 100, 200, 300, 400, 500, 600, 700, 800, 900, 1,000. Okay. We're only doing 1,000 today there. And this, a couple, there are a couple ... This may be 1,000, 2,000 left of levels of layers of that chakra. So you'll move up to the back of your heart. Can you see the green colour of the heart with it's petals?

Michelle: Yeah.

Dominic: ... And the inside everything else. See that one is not quite straight, so I used the energies to repair it. You can see the colours coming through to come and repair and the ... Now you're going to start turning over the petals, opening up the chakras. Here it is, we opened up one. We're now going to work with, um, opening up the levels. Including legal levels. 100, 200, 300, 400, 500, 600, 700, 800, 900, 1,000. 100, 200, 300, 400, 500, 600, 700, 800, 900, 1,000. 1,100, 2, 3, 4 ... And we're back to whatever we'd remain with. We did about 2,000 layers there at least. 2,200, 2,500 ... We're about 150 layers left but I don't want you to do the rest of them today. Um, so we're going to wait for the enlighten that chakra. We're now moving up to your ... We're moving up to your crown which has 1,000 petals. Can you see the pink type of look to the petals?

Michelle: Yeah.

Dominic: Okay. We're going to open up, um, this chakra as well. Here we go. Can you see it opening up?

Michelle: Mm-hmm (affirmative).

Dominic: It's opening up. And now we're going to work with the petals of, the nice little patterns inside of them swirling around and moving around, and you might see even a phoenix coming up to help you clean your chakras, to help you clean. But we'll, we'll investigate this relationship you have with this phoenix energy as a totem or as a enlightenment companion. In our sessions after next week when we start working with the karma, we will start working more with the legal level (pico scale) Okay. So here we go. And that's all that we want to do for today. Because that's, well, just about 6,000 layers, or a little bit more this week. So we're not going to move on. We've got about 20 minutes left of, of our time together. We're going to move on to working with opening up your (legal) listening.

Okay, so I'm going to sing a note on the spiritual level. You have a little bit of a legal listening, you have a little bit of lower astral listening. You don't have a lot of mid-astral listening or upper astral listening, but we're trying to get you causal listening so that you can, you don't have to have the upper mental or the astral. We're trying to get your causal listening so that you can hear the guides on the library, and you can listen to what happens in life books, listen to pictures, and you can listen to hear what's happening in life books, what's happening in ... To talk to the guides when they're in their home, and eventually when we start work, when we project to your library body we project like, astrally project, you can project into your library with sixth dimensional body. Explore the library, look at pasts, futures, dreams, relationships with ...

What we're going to start doing is exploring relationships with other beings, and you'll go into a life, come out of it, re, re, forgive, repent, forgive, um, apologise. And you'll meet beings then, we'll do constellations where you'll see your guides and your, uh, people from the past come up and we, you can actually physically, see them in your mind's eye, and we work with them in a constellation session. Or you'll look at your futures, you can walk around in and eventually when you have, um ...

If you, if you wish, you know, you can take classes with the school outside the library in psychic self-defence, healing, mechanics, angel geometry, there are about 8000 classes. You know they've got one for gardening and herbs, and one for, um, even like crating ice sculptures, and practical stuff like, um, for the guides and angels, how do you do dreams, how to repair life books, how to do a lot of different things, how to use a toaster and an oven, they have classes in that as well. Um, there are a number of different things that you can do there. Um ... And when the time comes, when you complete your process with your karma it will obviously activate your listening and you're going to a golden state of unity consciousness. lift you into this state. When you get closer to it your halo will come down and lift you into the state so that you can see the state of unity consciousness and, um, express a connection and see what duties you have.

That's probably when telepathy will start to happen for you, if not before then, should you choose to want to access that listening at that level. But that's something we can choose later on. We've got about 20 minutes so let's start with ... If you just want to listen to this note, and I want to see if you have, if you hear vibration or a, a sound in your head. Do you hear any vibration in any of your ears? We're starting with your ... With your left ear. Can you hear any vibration or feel, hear any ringing or chanting in your left ear?

Michelle: I don't know how intense it's supposed to be. I hear very light, a very, um quiet ... I ... It's almost like when you're getting a hearing test, you know?

Dominic: That's it.

Michelle: It's very, um ...

Dominic: It's going to start like that and then you're going to hear it more and more. It's going to be more pronounced and you'll eventually hear it at 100%. Okay, you've got about two or three guides I'm talking to to establish who's going to help with your, get, growing your listening over the next week so that by the next week when I activate your world of dreams and we open up the causal level, the records, you can start to talk and interact and hear what's going on. If at first you see, just see things, you can get subtitles in ... But I'd very much like for the, the causal listening to be in this. So I'm going to start singing that note again. It's quite a high note, it's like, you know, the sound is very high in the, high in the piano. It's like a, one of the top F's on the piano, or an E, an F or an E on top of the piano.

Can you hear it getting stronger?

Michelle: Yes.

Dominic: Okay. So.

Is the sound now a lot louder?

Michelle: Yes.

Dominic: Can you hear any voices? Any talking? Any music?

Michelle: No, it's, it's like ... It's like a frequency.

Dominic: Like (sings note) or like a frequency (sings note). [crosstalk]

Michelle: Like ... It's like when you ... I haven't done this in a long time, but get a hearing test and they say, "Do you hear this?" And they do something, it's like an electrical frequency.

Dominic: Mm-hmm (affirmative).

Michelle: That kind of ... And I'm not hearing words or, um ... [crosstalk]

Dominic: You feel the vibration. Are you hearing a note? A very high-pitched note?

Michelle: Yes.

Dominic: Okay.

Michelle: Yes.

Dominic: And I'm stopping now. And then I'm going to start again. Can you hear me that I've stopped, and then I start?

Michelle: Okay.

Dominic: Okay I'm going to stop. Does it stop then?

Michelle: Yes.

Dominic: Okay. Now I'm going to start it again and we're going to upgrade the percent. About 50, 52% at the moment in the left ear. Now working with your right ear to get it to about 50%, and then next week we will continue with the listening, and we will continue, um ... We'll activate your world of dreams and we'll show you around the causal, the akashic records, introduce you to that plane of existence and visit your guides. You can have a look around, look at the ... Work with me with the and we can look at, we can look at, you know, Jesus's life or Mary Magdalene's life, or Muhammad's life. Or you can look at the futures. Do, learn how to send a dream to yourself, you can learn how to repair a life book. There are a couple of things you can do in that level as well. Can you feel it opening up on the right?

Michelle: Yes.

Dominic: This one's got more listening on that level than the other ear. It's about, about 50/50 now. Checking both ears now, I want to see what you hear. The first sound, a much fuller sound than in the beginning.

Michelle: Mm-hmm (affirmative).

Dominic: Okay. Lovely. That's it for this week. If we want ... Can you make some time to see me next week?

Michelle: Okay. Um ... Yeah.

Dominic: What time suits you? Same time as we had this week?

Michelle: Um, yes.

Dominic: Lovely. I will see you then.

Michelle: Okay. All right, thank you Dominic

Dominic: Namaste. All the best moving into the next week and we'll see how it goes and, um, what we get it up to in the following week.

Michelle: Okay.

Dominic: Blessings to you.

Michelle: Thank you.

Dominic: Bye now.

Michelle: Namaste. Bye-bye. (call hanging up)

-5 dimension (Femto scale)

Quantum foam (also referred to as space-time foam) is a concept in quantum mechanics devised by John Wheeler in 1955. The foam is conceptualized as the foundation of the fabric of the Universe.

Based on the uncertainty principles of quantum mechanics and the general theory of relativity, there is no reason that space-time needs to be fundamentally smooth. Instead, in a quantum theory of gravity, space-time would consist of many small, ever-changing regions in which space and time are not definite, but fluctuate in a foam-like manner. Treatable with Yellow and White Reiki Rays.

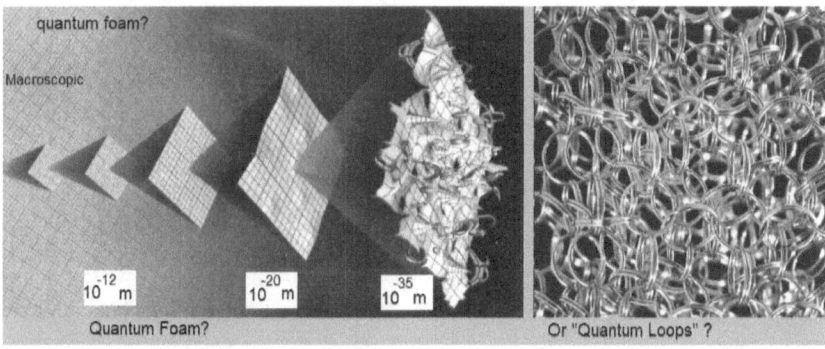

-6 dimension (Atto scale) Gnome

It is said in folklore that gnomes are one of a species of diminutive beings that inhabit the interior of the earth. They exist in the infra dimensions and act as guardians of its treasures. It is often seen when one relinquish relieve and resolve the Id through the mechanism covered earlier under how to use Negative-Dimensional Space. Use Red, Yellow and white Reiki.

What people often report seeing a tribe of gnomes and watch how boy children are born through a tunnel in the mountain. Those born with wings are born to id holders where an excess of ones children are girl children. In African cosmology this is the level where the natures helpers (water and earth sprites) lives.

-7 dimension (Pons planet)

The journey of beings start in the void that preceded creation where they nurse their hurts and find meaning in education as light energy moving though a Montessori shapes and colours matching game. As beings evolve through the movement of light they fall like comets to being trapped inside bubbles or omni-verses.

There is a way we carry this steward ship conscience throughout life. We sometimes ouch it in deep restorative sleep or moments of great meditation. When we reach near to this state the Pons level. We see a planet or a few cloud cities in the sky and forests on the planet surface below.

So wise parent do gardening in this realm at night to support the morals and virtues of their children. That they will later acquire as they age. Opening this state is a parental operation to open the grown up child to visual flips and auditory flicks.

[Use Grey and White Reiki Waves to cleanse this scale.]

-8 dimension (Sensei Flank)

Dominic: So last time. We met we were talking about. The emotional aspect your mental aspect and your training aspect. Reasoning your ability to predict. And work our long-term goals. Rather than short-term goals. And you know very So you've excelled at finding long term goals in 10 years and you were able to understand trends for 10 years but. You weren't doing this. [crosstalk] In the short term.

Dominic: We've had success with you with getting your stress down to a very small amount of stress which enabled you to make good decisions and get your management and get. Many deals done that would would have been very difficult. You went to many locations and you were able to handle many things. In your life .

Able to handle it with insight and expression and you didn't feel like you do feel that you were going to be forced to do anything and you feel fine. And even if you think by feel other people freaking you out any more and that's wonderful. So what's going on this last week what would you tell me about the week.?

Bruce: I managed to back get into training [at the gym] into personal time.

Dominic: Oh wonderful. Yeah.

Bruce: I manage to help at school and times. I'm managing to get up at 4:30am. Get to train till 5:30 feed the kids get to work at 8am. I'm trying no mater how bad Get to 6pm whether I finish or not. Yesterday I was just exhausted. I have lots of pressure by am less stressed by it.

Dominic: I'm glad you're getting back to your personal training.

That's very important. Okay. That's very important. You have this brute strength and you have wonderful. A wonderful star. That you're earning yourself. The gym was just one of the four. To make sure that you body is to stress free and that's wonderful.

Dominic: And tell me about your dreams. Are you having any dreams at night-time.

Bruce: Yes, a bit work related. Sometimes it has to do my training. So. It seems. That. Or on the social side. Things may. Be done. So. Quickly. This time.

Dominic: content we did say from the first session would involve. Like a night-time tournament or something about training and then it's interesting that your dreams been going around your lives. You've been giving insight into your life.

Dominic: About your social life your family and your work environment and through training which is one the. Option can be bad day-to-day things for the next three days around you could even last peoples day to make one day and the next day with a bit of a blur.

But sometimes it can also be productive and that's how you see your visual sense going. Its getting clearer of my time to see what's going on is something you see in my time. And as you advance with your sessions with me so you will advance with development of your inner senses and sensei strength. You will be able to preview the bank dreams before you have them. And you were in your instructor at the gym half of this which is really wonderful. At night-time in the daytime. And you're getting passed and you're getting better with your targeting and you can actually move. An opponent and you can also work with. The Chi energy. And you do energy the right opposes the body and you can use it to do many things which is wonderful. So did you moving your emotional target. And that's really wonderful. So.

From the dream.

Dominic: You remember a few of them which is really wonderful. I think that's great. That goes to show that you using your whole day in the gym sizing everything every single day. Then the maths is helping and it looks like the treatment is hardening. We able to sleep every single night last week.

Bruce: Yes.

Dominic: Wonderful.

Dominic: It's important to have images at night-time such as restorative sleep it's important to have dreams and REM cycles in your sleep to make it to work and to make sure that you are independent relaxed and it's not just static. Who's getting rest but you need to have slept. You need to have room cycles in this to make sure that you are active at night-time is not too dormant to. Sleep in that your brain is active in a different way than it usually is which is about creativity and it's about. How to Be creative. And it is very beautiful. When the brain is creative enough to them as you resolve the stress and you come up with ideas and then you really issues of the day.

Dominic: When you read through it. My time is eight hours at night which is a long time. To do it a studio planning and reading strategic planning very well. Night you will be able to be able to plan many weeks ahead and you'll be able to predict the people and work out what you should do.

Okay so there's this new night life. It's working well. Things are going well. It's exciting ask that you go out to talk to me about now with the shrub. It's only a different topic. It concerns you work leaving you think about. When you're at home. So. We're dealing. With issues. This. Is. This. Is. Leading. Me. To this. Week. It's not. The. Actual. Me. Doing this. This. Is me.

Bruce: Where I'm at at the moment mentally that I am not the business the business is me. Working on my business is getting me to a certain point. That's not my full potential. Working on me and how I interact with all the other businesses. And people. Mentally I crashed. Struggling with this this week. Getting things done in the normal working hours. So what. Is my point. I'm. Doing. It every day. We've. Moved. It. Gone home and by 8pm I was out.

Dominic: Consider that you've realized about. The business being their only goal it teens and you being separate from them. You have to run them as you know how to do that because that's part of your business. Acumen in that part of your training is part of your life.

Dominic: Can be you ancestry and different people in your family have to be talents with money and managing. And with. Vision and with. Forecasting and predicting trends. They were very good at this. It makes sense to me that you have a certain business style and your own has their own unique style. And it makes sense for you for these businesses you want to run them as an owner and that makes all its sense.

Dominic: Many of you will get to a point where they want to run the franchise the way you read it in a different way and there is more time every time. Our. Women want to do many things they were happy with the games they do. People will never work the same way and effectively as you do. So what I'm hearing is that you are. Needing to make it work possibly. Do you have the time to catch up with me.

Dominic: Time. Now that you realize that the businesses are functioning on their own. There are certain where their own money if one is acting like an S. They have a separate legal entity and have bank accounts they are doing fine. How in the world and do you see the difference in their habits and reputation which is different from yours. It's important to separate yourself from your business and your service to do otherwise.

Dominic: When a business gets bad news you get bad news inside your body. They will talk about the business. And sound if you like you're being assaulted or be. Emotionally by it. But it's tough being able to balance between being not attached. To the business and caring enough that you work everything in the business as if it were your son or daughter. How often have to find something is balance between both. That is a good thing.

Dominic: Okay. So.

Dominic: Have you got any more you from work from head office to bar the businesses in on me.

Bruce: Yeah. Its one of many. Give it. To me. If not you should be looking.

Dominic: So your basis is actually look if your business is operating like it's been business like it's a million and they have closed down and they're having to shovel away. Can you heard anything different. I would say the. Third year businesses are running according to natural. Cycles. Whether people want to buy in there or not they're going in there. But ultimately do your market research you'll see that their businesses have their own life cycle and seasons. And they also have it. And then the second is a certain Isn't that how is working. You can look to see it harder to make it happen.

Dominic: Or you can look at it because it's the section it's just up to the community and the people in the country the trade stock exchange and the many patterns on the map are changing the way these visits work but if they're consistently wrong every single year life a certain way and then this is unique going military people as far as with emerging the next shortfall in the next few months. If you see patterns like this happening to businesses you mustn't worry about it.

Dominic: As long as it doesn't get too out of hand. That's all there is. Now the interesting thing is that you managed to impress the people the last few months a few weeks ago managed to impress the boss and a regional manager and they were happy with you.

Dominic: So. I would like to work with. Measure your stress. You might feel like a tingle going down in this. It is 144 Kg's Stress is getting less. Expensive and even bigger it's getting less bad 10 kg's a week which is very means that. You're doing a huge meeting or your demand and your time period of time you solve. If doing this find these 10 kilograms you read. The body is getting used to doing this. which is wonderful.

Dominic: Okay. Okay.

Dominic: So. I'm going to do a session with you. You could choose two things out of two things. I'm going to talk about it. One is getting rid of the stress. which is very useful to have in a week. Time to get rid of the stress. You can get rid of stress you feel. Sleepy you feel recharged you will feel insightful and able to get through all your obstacles mentally and it's really important to have a stress. In this week or next week if there's time extra after the stress is gone. Once the stress probably be gone I will stop for me over.

Dominic: I want to talk about business. You've had in the past. That might be influencing your future.

Dominic: Because I think our work in cycles. Like there are chemicals in chemistry. There is balance. And they nearly killed me. Did you have with this and you have this and you have a credit score and you have so many things and variables that work out in financial terms. Where is he going to give you money is money. You don't know what these big losers is all these offers and. A hair dryer with their little fingers. Look at the statistics and they figure out who should they talk to.

So it is wonderful that you have insight to meet that person last time. One of the regional managers in your area and 80 percent of the debt. That you had. And because you were shown that you were on target and you were getting to places they allowed you to. Had to pay the balance. It's amazing you know asking this of everybody you ever see anybody who is not working to the maximum all they can do. Let's actually ask would you solve that and actually look at a person and as their whole life. Like a therapist this this then they and say person is doing all this they can do. But now for a quick Thai-Chi workout (Which I combine with a 30 minute Silver and Green Reki healing.)

Dominic: You prefer silence? Music? Okay here we go.

[played gentle music for 30 minutes]

Beyond the Omniverse

For some past this point is cloaked in Math and physics equitations for the rest is quantified in mythology. The world tree is represented as a colossal tree which supports the heavens, thereby connecting the heavens, the terrestrial world, and, through its roots, the underworld. Its units are subatomic quanta; Its motives divine.

There are 8000 sects within its religious root system. They tell a vibrant story from prehistory till now. This book is born of the political shenanigans that happen between their roots. The crossover of anthropomorphic 'gods' crossing over the root to join a new pantheon every few Aeons. Our history is its result. It all begins with a certain willow tree.

Specific world trees include village in Hungarian mythology, Ağaç Ana in Turkic mythology, Modun in Mongolian mythology, Yggdrasil (or Irminsul) in Germanic including Norse) mythology, the Oak in Slavic, Finnish and Balt mythodology, Kien-Mu or Jian-Mu in Chinese mythology, and in Hindu mythology the Ashvattha (a Sacred Fig)

Appendix 1

Mechanistic settings to align psyche component until number reaches 50:50. Full balance.

1. DNA – Calien / Sorgesbord Ring 40:60

2. PRIMORDIAL ATOMS – Mendalev / Konjerstein 50:50

3. EGO – Barnard Blex club 50:50

4. CHAKARA – Utter produces Kampala signet 50:50

5. MIND – Metlev Kornikov device 50:50

6. MOOD – Haarhoff – Barnett twin tub barometer 50:50

7. NADI / PAIN BODY – Haarhoff – Blake goggles50:50

8. SUPER EGO: - Cynthia & Bordman glove 50:50

9. TORUS – Haarhoff-Connerly math freight device 50:50

10. MUSE – Emma Jenkins – E.B. White volley ball 50:50

11. ID – Carlyle- Estronto MC Eccher frame 50:50

12. THANATOS – J. Netlev and Koen Frisbee net 50:50

Appendix 2

Bio-Electron photonics: Nano Animal totem guides

1. DNA – LEOPARD

2. PRIMORDIAL ATOMS - PENGUIN

3. EGO – APE

4. CHAKARA – ROOSTER

5. MIND – TIGER

6. MOOD – RHINO

7. NADI / PAIN BODY - GOOSE

8. SUPER EGO – ARMADELO

9. TORUS – GOAT

10. MUSE – FLAMINGO

11. ID – RAT

12. THANATOS – TORTOISE

References

1. Mann, Henry. Love Stream. The Essence of Love The Art of Becoming Fully Human (Kindle Locations 197-209). Coral Tree Publishing. Kindle Edition.

2. Katya Walter Phd. *Double Bubble Universe: The Layout.* Kairos Center, 02 Jun 2014

3. Luke Wolcott. *Imagining Negative-Dimensional Space.* Mathematics Department University of Washington. Jun 15, 2015

4. Sergey Uzunyan & David Hedinfrom *Leptoquarks: subatomic soul mates.* Northern Illinois University. Thursday, Oct. 15, 2009

5. David Abram. *The Spell of the Sensuous: Perception and Language in a More-Than-Human World*, Vintage, 1997

6. Burkert W (1996). *Creation of the Sacred: Tracks of Biology in Early Religions.* Harvard University Press. ISBN 978-0-674-17570-9.

7. Haycock DE (2011). *Being and Perceiving.* Manupod Press. ISBN 978-0-9569621-0-2.

8. Roys, Ralph L., The Book of Chilam Balam of Chumayel. Norman: University of Oklahoma Press 1967.

www.ingramcontent.com/pod-product-compliance
Lightning Source LLC
Chambersburg PA
CBHW030918180526
45163CB00002B/381